RANKING HAZARDOUS-WASTE SITES FOR REMEDIAL ACTION

Committee on Remedial Action
Priorities for Hazardous Waste Sites

Board on Environmental Studies
and Toxicology

Commission on Geosciences,
Environment, and Resources

National Research Council

NATIONAL ACADEMY PRESS
WASHINGTON, D.C. 1994

NATIONAL ACADEMY PRESS   2101 Constitution Ave., N.W. Washington, D.C. 20418

NOTICE: The project that is the subject of this report was approved by the Governing Board of the National Research Council, whose members are drawn from the councils of the National Academy of Sciences, the National Academy of Engineering, and the Institute of Medicine. The members of the committee responsible for the report were chosen for their special competencies and with regard for appropriate balance.

This report has been reviewed by a group other than the authors according to procedures approved by a Report Review Committee consisting of members of the National Academy of Sciences, the National Academy of Engineering, and the Institute of Medicine.

The National Academy of Sciences is a private, non-profit, self-perpetuating society of distinguished scholars engaged in scientific and engineering research, dedicated to the furtherance of science and technology and to their use for the general welfare. Upon the authority of the charter granted to it by the Congress in 1863, the Academy has a mandate that requires it to advise the federal government on scientific and technical matters. Dr. Bruce Alberts is president of the National Academy of Sciences.

The National Academy of Engineering was established in 1964, under the charter of the National Academy of Sciences, as a parallel organization of outstanding engineers. It is autonomous in its administration and in the selection of its members, sharing with the National Academy of Sciences the responsibility for advising the federal government. The National Academy of Engineering also sponsors engineering programs aimed at meeting national needs, encourages education and research, and recognizes the superior achievements of engineers. Dr. Robert M. White is president of the National Academy of Engineering.

The Institute of Medicine was established in 1970 by the National Academy of Sciences to secure the services of eminent members of appropriate professions in the examination of policy matters pertaining to the health of the public. The Institute acts under the responsibility given to the National Academy of Sciences by its congressional charter to be an adviser to the federal government and, upon its own initiative, to identify issues of medical care, research, and education. Dr. Kenneth I. Shine is president of the Institute of Medicine.

The National Research Council was organized by the National Academy of Sciences in 1916 to associate the broad community of science and technology with the Academy's purposes of furthering knowledge and advising the federal government. Functioning in accordance with general policies determined by the Academy, the Council has become the principal operating agency of both the National Academy of Sciences and the National Academy of Engineering in providing services to the government, the public, and the scientific and engineering communities. The Council is administered jointly by both Academies and the Institute of Medicine. Dr. Bruce Alberts and Dr. Robert M. White are chairman and vice chairman, respectively, of the National Research Council.

The project was supported by the Department of Defense, the Department of Energy, the Environmental Protection Agency, the American Petroleum Institute, the Monsanto Company, and the Coalition on Superfund. Contract #NAS90-191.

Library of Congress Catalog Card No. 94-66574
International Standard Book No. 0-309-05092-8

Additional copies of this book are available from the National Academy Press, 1-800-624-6242.

Cover art by Terry Parmelee. Parmelee, a Washington, D.C., artist, considers herself stylistically influenced by the Washington Color School of abstraction that was in its heydey when she was earning her MFA at The American University in 1967. Parmelee is represented by the Jane Haslem Gallery in Washington, D.C.

Copyright 1994 by the National Academy of Sciences. All rights reserved.
Printed in the United States of America

COMMITTEE ON REMEDIAL ACTION PRIORITIES FOR HAZARDOUS WASTE SITES

PERRY L. MCCARTY *(Chairman)*, Stanford University, Stanford, Calif.
YORAM COHEN *(Vice Chairman)*, University of California, Los Angeles, Calif.
MARK M. BASHOR, Agency for Toxic Substances and Disease Registry, Atlanta, Ga.
KIRK W. BROWN, Texas A&M University, College Station, Tex.
JAMES W. GILLETT, Cornell University, Ithaca, N.Y.
ALAN J. GOLDMAN, The Johns Hopkins University, Baltimore, Md.
MICHAEL R. GREENBERG, Rutgers University, New Brunswick, N.J.
ROBERT E. HAZEN, New Jersey Department of Environmental Protection and Energy, Trenton, N.J.
GLENN PAULSON, Illinois Institute of Technology, Chicago, Ill.
MITCHELL J. SMALL, Carnegie Mellon University, Pittsburgh, Pa.
LOUIS J. THIBODEAUX, Louisiana State University, Baton Rouge, La.
CURTIS C. TRAVIS, Oak Ridge National Laboratory, Oak Ridge, Tenn.
VICTORIA J. TSCHINKEL, Landers and Parsons, Tallahassee, Fla.
JULIAN WOLPERT, Princeton University, Princeton, N.J.
JEFFREY J. WONG, California Environmental Protection Agency, Sacramento, Calif.

Project Staff

RAYMOND A. WASSEL, Project Director and Program Director
ROBERT J. CROSSGROVE, Editor
ANNE M. SPRAGUE, Information Specialist
ADRIENNE L. DAVIS, Senior Project Assistant
RUTH P. DANOFF, Program Assistant

BOARD ON ENVIRONMENTAL STUDIES AND TOXICOLOGY

PAUL G. RISSER *(Chair)*, University of Miami, Oxford, Ohio
FREDERICK R. ANDERSON, Cadwalader, Wickersham & Taft, Washington, D.C.
MICHAEL J. BEAN, Environmental Defense Fund, Washington, D.C.
EULA BINGHAM, University of Cincinnati, Cincinnati, Ohio
EDWIN H. CLARK, Clean Sites, Inc., Alexandria, Va.
ALLAN H. CONNEY, Rutgers University, N.J.
JOHN L. EMMERSON, Eli Lilly & Company, Greenfield, Ind.
ROBERT C. FORNEY, Unionville, Pa.
ROBERT A. FROSCH, Harvard University, Cambridge, Mass.
KAI LEE, Williams College, Williamstown, Mass.
JANE LUBCHENCO, Oregon State University, Corvallis, Ore.
GORDON ORIANS, University of Washington, Seattle, Wash.
FRANK L. PARKER, Vanderbilt University, Nashville, Tenn., and Clemson University, Anderson, S.Car.
GEOFFREY PLACE, Hilton Head, S.Car.
DAVID P. RALL, Washington, D.C.
LESLIE A. REAL, Indiana University, Bloomington, Ind.
KRISTIN SHRADER-FRECHETTE, University of South Florida, Tampa, Fla.
GERALD VAN BELLE, University of Washington, Seattle, Wash.
BAILUS WALKER, JR., University of Oklahoma, Oklahoma City, Okla.

Staff Program Directors

JAMES J. REISA, Director
DAVID J. POLICANSKY, Associate Director and Program Director for Natural Resources and Applied Ecology
KULBIR BAKSHI, Program Director, Committee on Toxicology
GAIL CHARNLEY, Acting Program Director for Human Toxicology and Risk Assessment
LEE R. PAULSON, Program Director for Information Systems and Statistics
RAYMOND A. WASSEL, Program Director for Environmental Sciences and Engineering

COMMISSION ON GEOSCIENCES, ENVIRONMENT AND RESOURCES

M. GORDON WOLMAN (*Chair*), Johns Hopkins University, Baltimore, Md.
PATRICK R. ATKINS, Aluminum Company of America, Pittsburgh, Penn.
EDITH BROWN WEISS, Georgetown University Law Center, Washington, D.C.
EDWARD A. FRIEMAN, Scripps Institution of Oceanography, La Jolla, Calif.
W. BARCLAY KAMB, California Institute of Technology, Pasadena, Calif.
RAYMOND A. PRICE, Queen's University at Kingston, Canada
THOMAS C. SCHELLING, University of Maryland, College Park, Md.
ELLEN K. SILBERGELD, Environmental Defense Fund, Washington, D.C.
STEVEN M. STANLEY, Johns Hopkins University, Baltimore, Md.
VICTORIA J. TSCHINKEL, Landers and Parsons, Tallahassee, Fla.
WARREN WASHINGTON, National Center for Atmospheric Research, Boulder, Colo.

Commission Staff

STEPHEN RATTIEN, Executive Director
STEPHEN D. PARKER, Associate Executive Director
MORGAN GOPNIK, Assistant Executive Director
JEANETTE SPOON, Administrative Officer
SANDRA FITZPATRICK, Adminstrative Associate

OTHER RECENT REPORTS OF THE BOARD ON ENVIRONMENTAL STUDIES AND TOXICOLOGY

Science and Judgment in Risk Assessment (1994)
Health Effects of Ingested Fluoride (1993)
Pesticides in the Diets of Infants and Children (1993)
Issues in Risk Assessment (1993)
Setting Priorities for Land Conservation (1993)
Protecting Visibility in National Parks and Wilderness Areas 1993)
Biologic Markers in Immunotoxicology (1992)
Dolphins and the Tuna Industry (1992)
Environmental Neurotoxicology (1992)
Hazardous Materials on the Public Lands (1992)
Science and the National Parks (1992)
Animals as Sentinels of Environmental Health Hazards (1991)
Assessment of the U.S. Outer Continental Shelf Environmental Studies Program, Volumes I-IV (1991-1993)
Human Exposure Assessment for Airborne Pollutants (1991)
Monitoring Human Tissues for Toxic Substances (1991)
Rethinking the Ozone Problem in Urban and Regional Air Pollution (1991)
Decline of the Sea Turtles (1990)
Tracking Toxic Substances at Industrial Facilities (1990)
Biologic Markers in Pulmonary Toxicology (1989)
Biologic Markers in Reproductive Toxicology (1989)

*Copies of these reports may be ordered from
the National Academy Press
(800) 624-6242
(202) 334-3313*

PREFACE

The National Research Council established the Committee on Remedial Action Priorities for Hazardous Waste Sites in 1991. The committee was asked to examine the principal ranking methods being used or considered by federal and state agencies to rank hazardous-waste sites for remedial priority. Among the issues to be considered were the technical and policy purposes of the ranking methods, the effectiveness of the methods in achieving their intended purposes, the assumptions embodied within the methods, the uncertainties in the methods' results, and the methods' flexibility for considering new information and for analyzing and comparing the cost effectiveness of remediation. The project was supported by the U.S. Department of Defense (DOD), the U.S. Department of Energy (DOE), the U.S. Environmental Protection Agency (EPA), the American Petroleum Institute, Monsanto, and the Coalition on Superfund. In response to a request from DOD, an interim report—completed in June 1992—assessed the methods, assumptions, and constraints of DOD's Defense Priority Model (DPM), which was being developed to assist in decision making for hazardous-waste site restoration. This final report contains the results of a broader, more comprehensive study, not only of DOD's DPM, but also of EPA's Hazard Ranking System (HRS), DOE's proposed Environmental Restoration Priority System (ERPS), and to some extent, systems being used by various states.

Estimates of costs to meet the current national goals of cleaning up hazardous-waste sites extend beyond hundreds of billions of dollars. Serious questions are being raised as to whether the United States can afford to remediate all these sites. Regardless of whether priority setting is achieved explicitly in a manner open to public scrutiny, or implicitly in some obscure manner, priorities are certainly being set. A reliable system is needed to manage this expensive remedial effort in order to provide adequate protection of human health and the environment while making efficient use of financial resources. In this study, the committee examined the approaches being used by different agencies, assessed their effectiveness, and developed recommendations for the future.

During its initial meetings, the committee received information about the ranking models and overall priority-setting systems being used or developed by several federal and state agencies. The committee found no written descriptions of priority-setting systems that are used in the private sector to make remedial decisions. The committee was not asked to determine, nor did it conclude, whether any of the models was better than the others.

The committee considered the possible use of a unified national approach to improve the decision process for hazardous-waste site remediation in the future. Arguments for and against a unified procedure are presented in this report.

The committee would like to extend its appreciation for the cooperation provided by many individuals who furnished the committee with information about the different priority-setting systems. We especially would like to thank Marcia Read, Kevin Doxey, and Thomas Baca with the U.S. Department of Defense, and Judith Hushon with Environmental Resources Management Company, who devoted considerable time and effort to assist our review of the DPM and DOD's priority setting process. In addition, they participated in the scoring exercise of the five restoration sites. From the U.S. Environmental Protection Agency, we would especially like to thank Dorothy Canter, David Evans, Stephen Caldwell, and James McMaster for informing the committee about EPA's priority setting with the HRS. In addition, Lawrence Zaragoza of EPA assisted in the scoring exercise. From the U.S. Department of Energy, valuable input for the ERPS evaluation was provided by R. Patrick Whitfield, Thomas Longo, Frank Baxter, and Thomas Cotton (J. K. Associates). John Pendergrass of the Environmental Law Institute provided the committee with information on the numerous approaches being used in priority setting by various state agencies. Others who provided valuable assistance in the scoring exercise were Jill Morris (Oak Ridge National Laboratory), Michael Gresalfi (SAIC), and Stuart Haus (MITRE).

In such an effort, a great deal of coordination is required for committee meetings, much information must be gathered, and documents must be typed, distributed, integrated, and edited. For this, a great share of appreciation must go to Raymond Wassel, the NRC staff officer for this project. He not only expedited the interactions between committee members and provided valuable input to the report itself, but also consistently reminded the committee members of their duties and responsibilities. The committee also extends its appreciation to others on the BEST staff for their assistance, including James Reisa, director of the Board on Environmental Studies and Toxicology; William Lipscomb, research assistant; Felita Buckner, Adrienne Davis, and Ruth Danoff, Program Assistants; and Anne Sprague, information specialist. I would also like to thank others on the BEST staff who provided assistance for this effort and Robert J. Crossgrove, editor of this report.

Finally, I would like to thank the members of the committee, who devoted so much of their time to this effort. Their backgrounds are diverse and their perspectives about the issues under consideration varied considerably. They provided a stimulating environment for addressing the issues of importance. I believe the results of their efforts provide an excellent framework for priority setting in the future.

PERRY L. McCARTY
Chairman

CONTENTS

EXECUTIVE SUMMARY   1

1   HAZARDOUS-WASTE SITE PROBLEMS IN THE UNITED STATES   23

    Introduction and Charge to the Committee, 23
    U.S. Environmental Protection Agency Superfund Program, 25
    U.S. Departments of Defense and Energy Programs, 34
    Other Federal and State Programs, 39
    Setting Priorities, 46
    Discussion, 51
    Scope of the Report, 54

2   PRIORITY-SETTING PROCESSES   57

    Basis for Developing a Priority-Setting Approach, 57
    Desirable Features of a Priority-Setting System, 59

3   CLASSIFICATION OF PRIORITY-SETTING APPROACHES   65

    Introduction, 65
    Environmental Evaluation, 66
    Risk Analysis, 68
    Environmental Impact Analysis, 75
    Structured Value-Scoring Methods, 78
    Multiattribute Approaches, 80
    Cost-Benefit and Cost-Effectiveness Approaches, 82

| 4 | EPA's PRIORITY SETTING | 85 |

Background and History, 86
Role of the HRS in the Superfund Program, 89
Model Structure and Components, 93
The Effect of the 1990 HRS Revisions, 122
Priority Setting at Later Stages of Superfund, 125
Proposals for Improving Superfund Site Selection
   and Priority Setting, 126
Summary Evaluation on EPA Priority Setting
   for Hazardous-Waste Sites, 130

| 5 | DOD's PRIORITY SETTING | 135 |

Introduction, 135
Background and History, 136
The DPM Structure, 137
Pathway Scoring, 146
Contaminant-Hazard Scoring, 153
Receptor Scoring, 160
Socioeconomic Issues, 163
Scoring Methodology and Aggregation, 165
Validation, 167
Sensitivity and Uncertainty Analyses, 168
Summary, 172

| 6 | DOE's PRIORITY SETTING | 177 |

Background and History, 179
The Environmental Restoration Priority System, 182
Summary Evaluation of DOE's Priority Setting
   for Hazardous Waste Sites, 205

| 7 | STATE PRIORITY SETTING | 211 |

Introduction, 211
States with Ranking Systems Similar
  to the EPA Hazard-Ranking System, 212
Other Numeric Ranking Systems, 220
States with Ranking by Broad Category, 221
Discussion, 222

| 8 | COMPARING FEDERAL RANKING MODELS | 225 |

The Decision-Making Processes, 225
Comparative Scoring Exercise of
  Federal Ranking Models, 230
General Conclusions, 249

| 9 | TOWARD A UNIFIED NATIONAL APPROACH | 251 |

Advantages and Disadvantages of a
  Unified Approach, 252
Proposed Unified National Process
  for Setting Priorities, 259

| 10 | CONCLUSIONS AND RECOMMENDATIONS | 267 |

Need for Priority-Setting Process, 267
Current Priority-Setting, 268
Improving the Priority-Setting Process, 270
Current Ranking Models Used in Priority-Setting, 272
Improving the Models, 275

REFERENCES 279

Ranking Hazardous-Waste Sites for Remedial Action

Ranking Hazardous-Waste Sites for Remedial Action

EXECUTIVE SUMMARY

OVERVIEW

In 1980, Congress responded to public concern over the Love Canal situation by passing the Comprehensive Environmental Response, Compensation, and Liability Act (CERCLA). CERCLA initially established the $1.6 billion Superfund program to assess hazardous-waste sites, determine responsible parties, and provide expeditious financing for cleanups when responsible parties did not do so. CERCLA also required the U.S. Environmental Protection Agency (EPA) to develop a National Priority List (NPL) with a minimum of 400 sites for prompt cleanup. This mandate was extended in 1986 by the Superfund Amendments and Reauthorization Act (SARA), which added another $8.5 billion to finish the job. Implicit in both pieces of legislation was the idea that a few highly contaminated sites could be quickly identified and cleaned up. Experience proved otherwise.

Superfund has become a massive program. The number of sites requiring cleanup turned out to be far greater than originally anticipated. In 1977, a year before Love Canal and 3 years before CERCLA, EPA had reported on hazardous contamination at only 421 sites. EPA now expects the NPL to reach 2,000 sites, although other sources have estimated that the eventual total could reach 10,000.

Site remediation often has turned out to be far more complex than originally anticipated. For example, although a relatively small fraction of NPL sites is under the jurisdiction of the U.S. Department of Defense (DOD) and Department of Energy (DOE), those sites—where cleanup funding will be provided from the agencies' operating funds, not by Superfund—can be extremely complicated, including hundreds and even thousands of areas contaminated with large quantities and exotic mixtures of hazardous and radioactive contaminants. The closing of DOD bases and the decommissioning of DOE plants pose a number of additional social, economic, and political issues.

Site remediation is also proving to be far more expensive than originally anticipated. The original Superfund of $1.6 billion was designed to clean up 400 NPL sites at an average cost of $3.6 million per site; but by 1990, EPA was projecting a total cost of $27 billion at an average cost of $26 million per site. As new sites are added to the NPL, others have estimated that the total cost of Superfund alone could rise to between $100 billion and $500 billion over the next 30 to 50 years. When the DOD, DOE, state government, and private sector shares are added, the total bill for hazardous-waste site remediation could surpass $1 trillion.

This amount competes with expenditures for other pollution-control efforts, as well as other societal needs, such as reducing the national debt, providing health care, improving education, and

EXECUTIVE SUMMARY

renewing the nation's public works infrastructure. Given the many pressing national needs, it is doubtful that the United States will be able to fund the huge task of remediating all sites to the cleanest level possible. Thus, the agencies responsible for hazardous-waste site remediation—EPA, DOD, DOE, and others—will be required to make difficult and inevitably controversial choices. It no longer suffices to have a decisionmaking process or model that only attempts to identify the bad sites. There are too many of them. Faced with this reality, society needs a priority-setting system that helps define a systematic remediation strategy, addressing such questions as where and when available funds should be spent. The sheer cost of the enterprise—to the government, the taxpayer, and the U.S. economy—requires that priorities be set for waste-site remediation to protect human health and the environment.

CHARGE TO THE COMMITTEE

The National Research Council's Committee on Remedial Action Priorities for Hazardous Waste Sites was formed to assess the principal methods that federal and state agencies are using or developing to rank sites for remediation priority. The committee was asked to consider the intended technical and policy purposes and actual uses of the methods in the ranking decision process; their effectiveness in achieving those purposes; the types and levels of uncertainty of the input data and the methods' resulting limitations; the methods' assumptions; the appropriateness of the assumptions for the methods' intended purposes; the sources, magnitude, and treatment of significant uncertainties in each method; the sensitivity of the resulting score to the method's computation process; and the method's flexibility for follow-up evaluation of site assessments or for comparative analyses of the costs and effectiveness of remediation techniques. The committee was asked to identify the information and research needed to establish standards of performance and consistency for nationally applicable hazardous-waste site-ranking methods.

The committee examined the hazardous-waste site-ranking and priority-setting models developed or used by EPA, DOD, DOE, and some state governments to help them rank sites for remediation from among the tens of thousands of abandoned hazardous-waste sites. The committee also attempted to understand the larger processes by which these agencies choose sites to remediate and decide the level of remediation for each site.

Part of the committee's task was to prepare an interim report evaluating the methods, assumptions, and constraints of the Defense Priority Model (DPM), a ranking method developed by the U.S. Department of Defense. To meet this responsibility, in 1992, the committee completed an interim report entitled: *The Department of Defense Priority Model for Hazardous Waste Site Restoration: An Independent Assessment of Methods, Assumptions, and Constraints.* (The interim report is discussed in Chapter 5.)

DESIRABLE FEATURES OF A PRIORITY-SETTING SYSTEM

A priority-setting technique to aid in decision making for hazardous-waste site remediation should be consistent with the purpose intended. It should provide a formal, systematic, and consistent framework to catalog and compare information to help decision makers design strategies, allocate resources, evaluate progress, and inform the public. The factors to be considered include not only potential threats to human health and the environment,

EXECUTIVE SUMMARY

but also social and economic factors. A properly designed system should be comprehensive enough to address all of these factors, yet flexible enough to accommodate numerous, often competing objectives. Most important, the method by which information is obtained and used should be objective, explicit, and replicable, so as to preserve the credibility and acceptability of the larger priority-setting process. Ranking models that provide a framework for analyzing information and presenting results are often used as an important step in the process.

In this study, the committee found that less information was available about the overall priority-setting processes of the agencies than the ranking models used in that process. For much of the priority-setting processes, concrete procedural descriptions were typically unavailable, so that most of the committee's information had to be sought by exploratory questioning of agency officials and experts who met with the committee. As a result, the committee's findings focus more on *site-ranking methods* than the broader *priority-setting processes*. The distinction is an important one. Most of the systems developed to date are used only to rank sites according to some numerical score; these scores are considered, along with other factors, to arrive at actual remedial priorities, which can be quite different from the numerical scores.

With this caveat in mind, the committee attempted to compare the tools used or developed by EPA, DOD, and DOE for ranking sites as part of the priority-setting process. The committee evaluated the models based on their adherence to professionally accepted criteria for developing and applying site ranking models, which may be summarized as follows:

- acceptability and credibility, based on clear statements of the model's purpose and intended users as well as the incorporation of scientific peer review, public participation, and public comment during the model's development;

- adequacy of the model to account for the risks that contaminated sites pose to human health and the environment, as well as social and economic considerations;
- appropriateness of the model's logic and mathematical operations;
- adequacy of the model's documentation to explain and justify why it is designed as it is;
- thoroughness with which the model has been tested for validity (i.e., its ability to produce a reliable ranking of risks or sites); and
- appropriateness of uncertainty and sensitivity analyses (i.e., determination of uncertainties in model scores and their implications for site ranking and prioritization).

The committee also evaluated the extent to which the models exhibited other desirable features such as "transparency" (i.e., explicitness) and user-friendliness; flexibility in handling different waste sites and updated information; inclusion of cost estimates for remedial options; and security features to prevent unauthorized changes in site data, model parameters, and model outputs. Finally, the committee evaluated the results of a comparative assessment of the three major federal ranking models using a common set of input data from five contaminated waste sites.

PRIORITY SETTING AT EPA

EPA is involved in remedial decisions at many hazardous-waste sites and must deal with the whole gamut of interested parties and stakeholders. EPA must do all this under continuous public scrutiny and political pressure to remediate sites quickly. Since thorough risk assessments at tens of thousands of sites would be impracticable and unwarranted, EPA needs a mechanism to sort sites quickly on the basis of limited data.

EXECUTIVE SUMMARY

The first major step in the Superfund priority-setting process occurs when a nominated site is scored using the Hazard Ranking System (HRS) model. The HRS is a scoring system used to assess the relative threats associated with contaminant releases from different sites. The HRS combines various characteristics of the site, wastes, and surrounding environment to compute an overall score. As part of the calculations, separate scores are computed for each of four exposure pathways: groundwater, surface water, soil, and air. The HRS score, ranging from 0 to 100, is a screening mechanism for determining whether a proposed site is included on the Superfund NPL. Other scoring and ranking systems are used by EPA in later phases of the Superfund process, but such systems are considerably less formal than the HRS.

The committee judged the HRS model to be generally well documented and supported. Despite certain technical limitations (see Chapter 4), it is generally consistent with accepted scientific knowledge and has been subjected to extensive peer review, public participation, and public comment. The procedures for determining and combining HRS scores provide *relative* rankings of sites; it is consequently inappropriate to interpret the resulting HRS score in an *absolute* sense.

Results of the HRS model have been compared to the results of more detailed site assessments based on risk analysis and expert panels. The degree of HRS correlation with these estimates has generally been low to modest. In theory, because the model includes so many factor scores, it is relatively robust with respect to uncertainties in any one of them. In practice, however, the scoring outcome is quite sensitive to the overall effort exerted in data collection at the site, since the score for each environmental pathway is sensitive to the presence or absence of *observed* contamination in that pathway. Often, the more data, the higher the score. This creates the potential for misinterpretation or even manipulation.

The HRS model is broadly applicable to the types of hazardous-waste sites that EPA must evaluate. It emphasizes long-term risks, because EPA addresses immediate threats by other methods before HRS scoring. Because the scores are ordinal (i.e., relative), however, the HRS is inappropriate for selecting or tracking remedial actions. The model does not consider the costs or timing of remediation (issues that are considered at later stages of the priority-setting process), and it does not provide a basis to assess the relative weights given to human health versus ecological impacts.

Recent revisions to the original HRS have corrected some deficiencies but have also made the model significantly more difficult to understand and greatly increased the amount of time required to score each site. If EPA remains committed to an early decision about whether to list a site on the NPL, the HRS model, with appropriate modifications as recommended in Chapter 4, may remain the best alternative available. However, modifications to allow for more detailed review of sites with intermediate scores might help to reduce the number of sites which are not added to the NPL but would be if assessed more carefully, and sites that are currently included on the NPL but would not be if assessed more carefully.

PRIORITY SETTING AT DOD

In the past, DOD (and DOE) were not under great external pressure as they are today to clean up hazardous-waste sites. Many of their sites are in relatively remote or inaccessible areas, and national security considerations inhibited public scrutiny. The source of funds for these cleanups is the agencies' operating funds, not Superfund. As a result, DOD and DOE have had far more control over the choosing of sites for analysis and remediation, and this greater control is reflected in their priority-setting models.

EXECUTIVE SUMMARY

DOD's site-remediation goals are to remove imminent health threats, address the worst sites first, meet CERCLA and SARA requirements at NPL sites, and use resources effectively and efficiently. Unlike EPA, which uses the HRS model only for initial screening and NPL listing, DOD developed the Defense Priority Model (DPM) to assist in ranking sites for remedial action. Because DOD usually was under less pressure than EPA to make a quick yes-no decision at each site, it was able to obtain more field data for use in its model. The resulting numerical score, from 0 to 100, was intended to represent the relative potential threat that a contaminated site poses to human health and the environment.

DPM uses a combination of quantitative data and qualitative approximations. It calculates separate subscores for adverse effects on humans and ecological resources via surface water, groundwater, air, and soil pathways, and then combines them into an overall site score. The committee considers that to be a reasonable approach, but some of the assumptions, algorithms, and methods embedded in the model have a weak theoretical basis. DPM also does not explicitly address social and economic impacts, nor is it clear that DOD substantially addresses these factors through a separate evaluation process.

The DPM is user-friendly and its structure is clear, but it contains portions that have not been validated, and there has been no attempt to validate the overall model. Further, the DPM's linear scale produces a very tight range of site scores relative to the limits of 0 to 100; in 1991, 65 percent of the 284 sites evaluated with DPM had scores between 13 and 37. This narrow interval may limit DPM's ability to discriminate between sites. A simple sensitivity analysis of the 50 sites with the highest DPM scores demonstrated that uncertainties in model inputs and structure can have large effects on the scores and ranking of sites. Spreading out the numerical scores with alternate algorithms might allow better discrimination among sites. Because DPM was still undergoing development during the committee's evaluation, it has the potential for this kind of revision and improvement. (After completing its analyses, the committee was informally told that DOD has decided not to use the DPM for site ranking.)

Priority Setting at DOE

DOE faces two additional challenges in cleaning up its hazardous waste sites. The first is the greater complexity of the problems at sites that often contain radioactive materials, toxic chemical wastes, and mixed (radioactive-chemical) wastes. The second is that more than 60 percent of DOE's cleanup funds are committed by formal agreements with EPA regions and the states, mandating that certain sites be remediated by specific dates. Consequently, DOE must optimize the allocation of scarce resources subject to a series of constraints, notably these legal agreements.

To manage its complicated remedial effort, DOE has developed an Environmental Restoration Priority System (ERPS), the goals of which are to document and support DOE's budget requests and to allocate funds among its programs and installations. The system is more comprehensive than HRS or DPM, explicitly addressing social and economic impacts, cost considerations, and uncertainties. However, it does not use a "worst-first" approach, and the results would not necessarily lead to the remediation of sites according to the magnitude of the risks they pose to public health and the environment. ERPS builds more of the decision-making process into the model itself, rather than relying on an external process that may not be as amenable to outside scrutiny. ERPS is less than 5 years old, has not been applied much in the past, and is not expected to be used by DOE in the near future, so the committee was unable to evaluate it as thoroughly as it did the EPA and DOD models.

ERPS is based on a sophisticated implementation of multi-

Executive Summary

attribute utility theory, a decision-making approach that allows for the simultaneous evaluation of multiple objectives and factors—in this case, not only health risks but also environmental, social, economic, and political impacts. Implementation involves four steps:

- *Structure the decision problem* by specifying the objectives, identifying the alternatives, and determining attributes or outcome measures by which results can be assessed.
- *Assess the possible impacts of different alternatives,* using probability distributions if exact effects cannot be determined.
- *Determine the preferences (values) to the decision maker* by assigning different weights to the various attributes or effects of each alternative.
- *Evaluate and compare alternatives.*

DOE's computer model known as MEPAS (Multimedia Environmental Pollutant Assessment System) is used to generate summary risk indicators for use in ERPS. MEPAS focuses only on adverse health impacts, not on environmental or other effects; but it includes radioactive, chemical, and mixed wastes, multiple pathways, and both direct and indirect exposures.

The ERPS procedure for addressing health risks suffers from a lack of data as a basis for most of the estimates involved. The part of ERPS that combines individual risks with population risks appears to overemphasize large populations exposed to low risks, at the expense of small populations exposed to extremely high risks. A more sophisticated approach to address risk variability is needed.

One of the most innovative aspects of ERPS is "uncertainty reduction" as an objective in ranking budget cases. In the early stages of environmental restoration, a great deal of uncertainty may exist at an installation regarding the actual levels and types of wastes, the risks they pose to health and the environment, and the costs of mitigating these problems. In ERPS, formal approaches using decision-analysis techniques are directed toward determining, in explicit economic terms, the value of activities that will eliminate these uncertainties.

PRIORITY SETTING BY STATE GOVERNMENTS

State governments have important responsibilities for hazardous-waste site remediation. Sites placed on the NPL will be cleaned up with federal money only if the state agrees to pay 10 percent of the capital cost and all future operating and maintenance costs. Many states also have "state Superfund" programs for dealing with sites even if they are not placed on the federal NPL. Some states have made an investment in hazardous-waste site remediation that, on a per capita basis, rivals those of the federal agencies.

The Environmental Law Institute (ELI) reported that, as of 1991, 29 states were operating remediation programs supported by enforcement authorities and dedicated funds; another 12 states had legal authority to conduct cleanups but lacked funding and staffing. Several states have multiple statutes that provide authority for various remediation activities at sites that are not covered by Superfund. ELI found that 24 states had their own priority-setting systems. These states do the HRS scoring for EPA and typically choose as state Superfund sites all of those not forwarded to EPA for listing on the NPL. The committee observed many different scoring approaches among the states, although they generally used the same types of input data.

The state approaches examined by the committee fall into three categories: (1) HRS-like models (e.g., California, Ohio, Oregon, and Washington); (2) other explicit numeric systems leading to a site-specific score (e.g., Michigan); and (3) characterization of sites into three or more groups based on narrative description of the

Executive Summary

severity of effects (e.g., Missouri, Montana, and New York). The third approach evaluates up to seven characteristics but involves no mathematical combination of factors to yield a score. Any one of a number of potential effects could lead to a maximum score, and the analyst is given considerable flexibility in deciding which potential effects to pursue in more detail.

In many cases, there is evidence of very thoughtful development of site ranking models by state agencies. However, the relationships between the model parameters and the strategies for combining the parameter values are often unclear and undocumented.

Comparative Assessment of Federal Priority-Setting Models

The committee undertook an assessment of the three federal ranking models—HRS (EPA), DPM (DOD), and MEPAS (DOE)—to compare the outcomes they would yield from a common set of input data developed from a common set of five actual waste sites. The committee took into consideration that the agencies developed their models for different purposes and for use at different stages of their respective decisionmaking processes. Despite these differences, however, the three models should be expected to produce generally similar *relative* rankings of the same set of hazardous-waste sites—that is, they should generally give the same indication of which sites are "worse" (i.e., produce higher risks) compared with other sites. DOE's ERPS model was not included in the scoring exercise because its design and application are so different from the HRS, DPM, and MEPAS models.

Differences among these models make exact comparisons difficult. For example, DPM and HRS are scoring systems that assign points to a site based on important site characteristics, but without directly modeling the process of contaminant transport.

MEPAS, on the other hand, includes algorithms for contaminant migration and fate and the resulting risks to human health.

The committee provided each agency with a common set of site descriptions, narratives, background, and data for 5 sites. Each agency was asked to run its own ranking model and to provide the committee with the resulting scores. To ensure comparable applications, the three agencies communicated extensively during the exercise. After completing the model runs, each agency summarized its results, scaled its scores between 0 and 100, and submitted reports to the committee.

In general, the scores from the three models tended to follow similar trends from site to site, but in each model, the reasons for higher scores were often very different. Even when the models produced generally consistent site scores, they tended to differ as to which were the dominant risk-producing contaminants and transport pathways. The three models did not all agree on which of the five sites posed the highest potential risk. The dominant factor leading to differences in site scores among the models appeared to be differences in the selection and weighting of site data, particularly for contaminants and environmental pathways, and not differences in model structure.

Overall Conclusions and Recommendations

Current Ranking Models

The committee believes that formal mathematical ranking models play less of a role than they can and should in determining remedial priorities for hazardous-waste sites. However, all of the current ranking models were found to fall short on several important aspects of model development, including documentation, validation, completeness, transparency, and inclusion of social and economic factors. There is a large base of scientific knowledge on which to build such a system, and there are no large gaps that would first require further research, but the agencies currently use this science base differently. The mixture of science and policy components in the current models, while not inappropriate, complicates analysis and comparisons among priority-setting processes.

The committee recommends that the agencies work in close collaboration to improve their model development programs in at least four general ways:

• clearer documentation of core elements of the model for technical and lay audiences;
• greater public involvement in the process of developing a model and applying it to a given site;
• better validation of model components, reference data, and parameter values to reflect current and new knowledge; and
• more explicit consideration of social and economic factors, particularly the costs of remediation alternatives.

Current Priority-Setting Processes

The site-ranking models are only one part of the overall priority-setting process for hazardous-waste site remediation. By comparison, the agencies' current priority-setting processes themselves are not well defined, and appear to lack adequate evaluation, consistency, and effective oversight. None of the agencies defines its overall priority-setting process in a manner that is explicit, clear, well documented, and open to scientific and public scrutiny. Approaches to priority setting are not always consistent even within a given agency, and there is no consolidated priority setting process for sites at the national level.

Toward a Unified National Approach

The enormous costs of environmental remediation certainly justify the development of an objective, replicable, and equitable priority-setting process that is fully open to public scrutiny. At the present time, there is no consistent relationship between the hazard present at a site and the process by which the different agencies screen and evaluate a site for remediation. For example, EPA works closely with DOE and the states to develop remediation plans for some DOE sites, but other sites are the responsibility of a single agency, and each agency has developed its own unique protocols. As a result, it is extremely difficult to compare the degrees of cleanup and levels of protection being provided or even pursued by the different agencies.

The process of evaluating and cleaning up sites no longer follows the simple model that the creators of Superfund envisioned. During a period of pressure to accomplish more with fewer resources, the use of several independent and inconsistent methods may be neither effective nor prudent. The committee therefore recommends that the government consider the development of a unified national process of scientific hazardous-waste site analysis to replace the current multiple approaches. Specifically, the committee recommends a new system designed to achieve three main goals:

- *Greater consultation and collaboration among the agencies.* This goal is the least intrusive to existing agency approaches. EPA, DOD, and DOE should form an interagency task force to coordinate the use of site-ranking models and determine how they can share data, expertise, quality control, validation procedures, and other information. The agencies would also share their respective approaches for including social and economic factors and for communicating with interested parties.
- *Scientific consistency.* This would require that each site be

EXECUTIVE SUMMARY

subjected to the same scientific protocols for evaluating health and safety, environmental impact, and economic costs and benefits. A unified approach could result in more explicit, thorough, and credible scientific input into the political process of deciding on resource allocation for site identification, ranking, and remediation.

• *Decision-making consistency.* This would add procedural and geographical consistency to scientific consistency. That is, all agencies would apply the same decisionmaking protocols to every site. Priority-setting decisions would not be influenced by which agency was responsible for the site—solvent spills at a factory in Illinois would be treated the same way as solvent spills at a DOD base in Arizona or a DOE facility in Ohio. Priority could even be assigned by a central interagency group, not exclusively by the parties currently charged with remediating the site. Decision-making consistency could require some reorganization of authority among federal agencies and would probably include some shifting of funds among agencies.

The committee recommends a three-tiered approach. This approach draws heavily on procedures already being used by federal agencies, either explicitly or implicitly, and thus requires no radical changes in thinking or development. Mathematical models would be used in all three tiers to assist in ranking sites. However, some factors are not readily quantifiable and thus would need to be addressed outside of the models.

The first tier involves *screening* candidate hazardous-waste sites. The site is evaluated to determine whether to (1) move it to the second tier for detailed characterization, (2) eliminate it from further consideration, or (3) gather more data before making a decision. These decisions would be based on limited information about potential risks to human health and the environment. EPA's HRS model and DOD's Defense Priority Model were developed, in part, with such a purpose in mind.

The second tier involves detailed *site investigation* to assess the extent of contamination at each site, the various environmental media and populations that may be affected, and the costs of remedial actions. The data gathered here should be sufficient to conduct formalized assessments of relative risks to human health and the environment, including the rates at which contaminants spread in time and space. In addition, at least preliminary estimates should be made of the economic damages to the resources. The overall objective in the second tier is a relative ranking of sites and a cost evaluation for at least three alternate levels of remediation:

- remediation that is sufficient to contain hazardous contaminants so they no longer present a significant risk to human health or the environment (a no-action alternative might be appropriate for this level at some sites, but at other sites land-use controls and restricted access might be required);
- remediation that is sufficient to restore the site to the point where no land-use restrictions are necessary, and
- more extensive cleanup (comparable to returning the site to precontamination quality).

The third tier would combine the ranking by risks and the estimated costs of remediation alternatives to determine what sites to address first and what levels of control to pursue. This process would still involve some mathematical formalization, but it would also include broader social and economic considerations that would be addressed outside of a mathematical model. This process needs to be more explicit than current practices so that funds can be allocated in a more open and cost-effective manner. The committee does not recommend a particular framework for doing this, but clearly one is needed.

A unified national approach would have the advantage of being compatible with current agency practices. It could make use of knowledge gained from the application of current models. Be-

EXECUTIVE SUMMARY

cause the agencies would all use the same procedures in Tiers 1 and 2, they would be able to devote greater effort in collaboration to examine the scientific basis for the mathematical algorithms, evaluate the validity of the models with respect to their intended use, and determine the sensitivity of the models to data inputs. The agencies could also share the costs of developing documentation and acquiring appropriate data inputs. The resulting consistency could increase the overall credibility of the process.

A single consistent national process that explicitly includes calculation of economic costs and benefits of remediation would be advantageous to decision-makers because it would make more explicit the reality that costs and benefits are always factored in some way into decisions. Such an explicit method would increase the credibility of the process by providing estimates that could be compared with actual site costs and benefits—in a sense, a kind of costs-and-benefits accounting. Also, a clearly documented costs-and-benefits protocol greatly decreases the chances of inadvertent or intended skewing of cost and benefit estimates for reasons that have nothing to do with hazard or remediation outcomes.

The three levels of remediation costs considered in Tier 2 would allow better judgments to be made concerning the degree of remediation to pursue at a given site. The relative costs for different levels of remediation would be provided more explicitly, in a manner understandable by decision makers and the public. For example, if the estimated cost of remediation at a given site (Site One) is $2 million for the first level and $3 million for the second, while the costs at a second site (Site Two) are $2 million and $100 million, respectively, and if both sites pose roughly equivalent risks, it should generally be convincing that the best use of limited remediation funds would likely be to clean Site One to the second level and Site Two to the first level. The benefits of cleanup would also be calculated. The decision for the two sites in the example might change if the benefits of the second level remediation of Site Two were $400 million compared to $100 million in costs.

19

When funding shortfalls occur, a single consistent process could facilitate negotiations among federal agencies, state and local governments, tribal governments and Native American organizations, private organizations, citizens groups, and other interested parties because they would all be negotiating from the same data base and on a more level playing field. Ideally, such a uniform approach would help the federal agencies develop a joint strategic plan for remediation under a variety of resource-constrained scenarios.

Another advantage of a single uniform scientific approach is that every analysis would treat every person the same and every forest consistently, regardless of whether they are located in an urban area of New Jersey or Louisiana, or a rural area of Maine or Arizona. The decision-making would allocate remediation resources on the basis of costs, benefits, and need for remediation rather than on the basis of the ability of a responsible party, state, business, or federal agency to pay.

The committee believes that a unified approach to setting priorities would better accommodate changes in the scientific, technological, economic, and political processes in the United States and abroad than do the existing multi-organizational approaches. This approach would provide a more rational basis for decisions about priority setting and levels of remediation at hazardous-waste sites.

Vast resources will be allocated for hazardous-waste site remediation throughout the 1990s and beyond. If the United States is ever to adopt a uniform national scientific and decision-making process, it makes sense to do it soon.

Ranking Hazardous-Waste Sites for Remedial Action

1

HAZARDOUS-WASTE SITE PROBLEMS IN THE UNITED STATES

INTRODUCTION AND CHARGE TO THE COMMITTEE

This report discusses the ranking methods and overall priority-setting approaches used by or developed for the U.S. Environmental Protection Agency (EPA), the U.S. Department of Defense (DOD), the U.S. Department of Energy (DOE), and some state governments to choose sites for remediation among the tens of thousands of hazardous-waste sites. As complex components of the priority-setting approaches, the ranking methods combine available information about waste sources; air, water, and soil pathways for contaminants; toxicants; and population and resources at risk to attempt to produce integrated numerical values. The resulting rankings can be used together with other social, economic and political factors to set priorities for cleanup. Reflecting the complexity of the ranking procedures, much of the report addresses technical issues such as accuracy of toxicity data bases, appropriateness of mathematical operations performed and applicability to various kinds of waste sites.

The National Research Council's Committee on Remedial Action Priorities for Hazardous-Waste Sites was formed to assess the principal methods that federal and state agencies are using or developing to rank sites for remedial priority. The committee was asked to consider, as part of its analysis, issues such as the following, and others that it considers relevant:

(a) the intended technical and policy purposes and actual use of the method in the ranking-decision process, and its effectiveness in achieving the intended purposes;

(b) the types and levels of uncertainty of input data resulting in the method's limitations;

(c) the method's assumptions (explicit and implicit) and the appropriateness of the assumptions for the method's intended purpose;

(d) source, magnitude, and treatment of significant uncertainties in each method;

(e) the sensitivity of the resulting score to the method's computation process;

(f) the method's flexibility for follow-up evaluation of site assessments (e.g., regarding changes in risk) or for comparative analyses of the costs and effectiveness of remediation techniques.

The committee was asked to make recommendations regarding information and research needed to establish standards of performance and consistency for nationally applicable ranking methods for hazardous-waste sites and to provide a basis for refining existing methods to improve the decision process for hazardous-waste site management and remediation in the future.

As background before addressing the scientific content of the methods, this chapter briefly reviews the recent history of the issue. Hazardous-waste site-remediation programs began with the goal of quickly cleaning up a limited number of highly contaminated and highly politicized sites attributed to American industry. As will be seen, however, these programs have evolved into endeavors whose full implementation in their present forms might not be possible because of technical limitations and costs.

The chapter addresses the EPA Superfund program, the DOD and DOE remediation programs, and other federal and state programs. Their mandates, program sizes, and cleanup costs are addressed. Priority setting is then discussed. In short, this chapter tries to answer two questions: (1) Why is the hazardous-waste site-management problem so much more complex than it was perceived to be a decade ago? (2) What role might a science-based ranking system play in the overall priority-setting process, considering the increasing political, legal, and economic pressures being placed on decision makers?

This chapter is not intended to present a comprehensive enumeration of all the legal mandates and political forces. (Some additional details are provided in Chapters 4-6 and in the citations therein.) Rather this chapter focuses on the programs that set the tone for hazardous-waste site remediation in the United States. The committee has tried to present an objective and balanced sample of the issues.

U.S. ENVIRONMENTAL PROTECTION AGENCY SUPERFUND PROGRAM

EPA is directly responsible for or indirectly involved in hazardous-aste cleanups of private and government sites. It is the agency of last resort if no party assumes responsibility for site cleanup. Some sites no longer operate; some are still operating;

and some facilities have both operating and closed sites. This section focuses on the Superfund program because of its historical, legal, and symbolic importance. Other EPA responsibilities are briefly reviewed elsewhere in the chapter.

Initial Mandate and Funding

In 1980, Congress mandated EPA to cleanup abandoned hazhazardous-ardous waste sites. The Comprehensive Environmental Response, Compensation, and Liability Act (CERCLA) of 1980 (P.L. 96-510) established a $1.6 billion "Superfund" to identify sites contaminated by hazardous waste, to determine responsible parties, and to finance cleanups when responsible parties could not. EPA was required to develop a National Priority List (NPL) with a minimum of 400 sites for cleanup. At least one site from each state had to be on the NPL. Superfund was extended for 5 years in 1986 by the Superfund Amendments and Reauthorization Act (SARA)(P.L. 99-499), and another $8.5 billion was added to finish the job. In addition, EPA was given guidance about additional risks to consider, research needs, modifications to its hazard-ranking system, and relationships with states and with the U.S. Agency for Toxic Substances and Disease Registry (ATSDR). (Some of that guidance is reviewed in this chapter. Chapter 4 provides a detailed presentation of priority setting within EPA's Superfund program.)

Original Hazard-Ranking System

EPA defines a hazardous-waste site in terms of the risks presented to human health and the environment at a specific location. The extent of the site depends upon the extent of the contamination. As assessment and remediation move forward, a site overseen by EPA is likely to be split into "operable units" that each require a specific remediation. Later, once a Superfund Record of Decision is filed and remediation has occurred, any natural resource damage must be reevaluated. For that purpose, a variety of sites or operable units can be combined for consideration by the Natural Resources Trustees.

CERCLA required the identification of 400 sites for an NPL, but provided very limited guidance for a hazardous-site ranking system. Section 105(8)A mentioned the following as relevant factors: population at risk, potential for drinking-water contamination, direct human contact, and destruction of sensitive environments. EPA retained the MITRE Corporation to develop a method for ranking and choosing 400 NPL sites. Because the Hazard Ranking System (HRS) that MITRE developed has had a major impact on priority-setting approaches used by other government agencies, a brief review of the original HRS (Chang et al., 1981) follows; a detailed presentation of the revised HRS is provided in Chapter 4.

MITRE's scientists reviewed existing methods of rating sites' relative hazards. They concluded that the existing models focused on water pollution impact, had no air pollution, soil pollution, or direct human contact elements, and had no method of integrating the different impacts (Chang et al., 1981). The need to meet CERCLA's mandate clearly demanded a state-of-the-art advance in priority-setting models. The original HRS expanded consideration of hazardous-waste site impacts to five components: groundwater, surface water, air, fire and explosion, and direct contact. The model required data on waste characteristics, quantities, releases, and targets for each of the five components for each site; thus the original HRS provided a method of integrating site-related data into a single site score (Chang et al., 1981).

In 1981, a decade before this National Research Council committee was convened to study priority-setting methods, the

MITRE staff identified limitations of their work that remain as unresolved issues today (see Chapter 4 for greater detail). The model is not amenable to an economic analysis because the hazard-ranking scores are only relative on a scale from 1 to 100. A site score of 60 is not twice as bad as one of 30. Second, quality control of the data and sensitivity tests of the results (with respect to data and model uncertainties) were limited. Third, the model did not take into account socioeconomic factors; these were to be evaluated by EPA after the model had been used to rank sites.

The required 400 sites were selected by using the original HRS. The cutoff score for the top 400 sites turned out to be 28.5. With only slight modifications described in Chapter 4, that cutoff score has continued to be the criterion used to determine if a site should be placed on the NPL. The use of a scoring model such as the HRS to dichotomize sites has been criticized (OTA, 1989; Hird, 1990). The major concern is that adequate procedures be used to ensure that dichotomizing sites will not exclude some sites that should be on the NPL.

Program Size and Cost: EPA

Superfund has become a massive program. One reason is that the number of sites that require cleanup has proved to be much larger than the initial estimates. In 1977, 1 year before Love Canal made national headlines and 3 years before the passage of the Superfund, EPA reported findings of hazardous conditions at 421 disposal sites (EPA, 1977). The principle problem was clearly chronic pollution of water, not imminence of catastrophic explosions and fires. Groundwater pollution was observed at 61% of the sites and surface water pollution at 40%. Fires and explosions were reported at 4%

One year later, after the identification of contamination at Love

Canal, Thomas Jorling, then EPA's assistant administrator for water and waste management, addressed the concern that government might be called upon to fund other Love Canal-type cleanups by requesting that the ten EPA regional administrators estimate the numbers of hazardous-waste sites and those that might be imminent hazards (Greenberg and Anderson, 1984). The response was 32,000 sites, of which 838 that might be imminent hazards. The estimating methods were, however, imprecise and nonuniform across the regions. In 1979, Fred C. Hart Associates, under contract to EPA, used a more systematic method to estimate the number of hazardous-waste sites at 50,664, of which 2,027 might pose a significant threat (Greenberg and Anderson, 1984).

After CERCLA was passed, EPA began to keep records of the number of sites identified as hazardous or potentially hazardous. In January 1983, for example, 13,392 such sites had been reported (Greenberg and Anderson, 1984). By June 1986, the year Superfund was reauthorized, EPA had 24,269 sites in its inventory with 951 on the NPL (Conservation Foundation, 1987). That is, in 3 years the number of identified sites had increased over 80% and the number of priority sites already was more than twice the original 400 required by Superfund legislation. As of April 1994, EPA reported 37,987 sites in its CERCLIS (CERCLA Information System) inventory of potentially hazardous sites.

The number of sites will almost surely continue to increase. The question is how many more will be added to the NPL. EPA expects to add about 100 new sites a year for the foreseeable future (S. Caldwell, EPA, pers. comm., January 1992). EPA (1990a) estimated that the number of NPL sites could reach 2,000. The Congressional Office of Technology Assessment (OTA, 1985, 1989) expects there will be 4,000 NPL sites by the year 2000 and that the NPL might eventually contain 10,000 sites (EPA's estimate of 2,000 sites; another 1,000 currently active hazardous-waste sites; 5,000 solid waste sites; and 2,000 sites to be found because of better site identification). Russell et al. (1991) predicted a range of 2,100 to 6,000 NPL sites.

Initial underestimation of the average cost of cleaning up sites is another major reason why the Superfund program now appears so much more costly. Enacted on December 11, 1980, the $1.6 billion trust fund was to come from taxes imposed on oil (raw materials) and on 42 specific compounds. About 13% was to come from general revenues (Superfund Fact Sheet, 1981). This formula was a matter of concern for the chemical industry, but appeared to be a manageable way of collecting $1.6 billion.

Speaking in early 1981, Senator Robert Stafford (1981) estimated the average cost of cleaning up a site at $3.6 million; which would make the Superfund's $1.6 billion enough to cleanup the stipulated 400. In 1990, EPA reported an average cost of $26 million to cleanup an NPL site, yielding a projection of $27 billion to cleanup existing NPL sites (EPA, 1990a,b). The estimate excludes costs for remediating sites not listed at the end of fiscal year 1990.

Other estimates place the cost closer to $100 billion or more for the EPA program. OTA (OTA, 1989; Passell, 1991) estimated the cost of cleaning up 4,000 NPL sites to be $80 to 120 billion, and a potential cost of $500 billion spread out over 30 to 50 years to cleanup 10,000 sites without major changes in the program goals and technological innovations.

The "best guess" estimate of Russell et al. (1991) is that $151 billion will be spent from 1990 to 2020 to remediate 3,000 nonfederal NPL sites, at an average site cleanup cost of $5 to 15 million. The general picture emerging from these more recent projections is that Superfund cleanup will cost the U.S. 10 to 50 times the amount of money allocated to EPA under CERCLA and its reauthorization. The Congressional Budget Office (CBO) reported a base-case estimate of $75 billion to clean up Superfund sites from fiscal year 1993 onward (CBO, 1994). A low-case estimate of $42 billion and a high-case estimate of $120 billion were also provided. CBO indicated that its estimates are lower than comparable EPA and Russell et al. (1991) estimates, primarily because of the assumptions about the future incidence of mega-sites and the costs saved in private-sector cleanups.

The Setting of Cleanup Goals

Another aspect of the Superfund program that bears upon priority setting is the slow pace of cleanups. This is related to a definition of a clean site, a matter on which the initial legislation, CERCLA, offered no guidance. Initially, EPA generally accepted containment of pollutants at sites or removal and transportation of wastes to permitted landfills. However, paving over wastes and building clay barriers around them failed to conciliate residents who wanted the wastes removed, not contained. Removing and transporting wastes may satisfy residents near a site, but other interest groups often objected to transporting wastes from one town to another (OTA, 1985). Furthermore, some local and national interest groups were not satisfied with EPA's ten regional administrators, who typically decided the extent of cleanup required at a site. That permitted outcomes that seemed inequitable to the groups—sites with residual cancer risks of 1 in 10,000 in one region would not be declared high risk, while the same site might have been considered a major risk in another region (Harris and Wrenn, 1988). In 1986, SARA provided specific guidance that had not been in CERCLA. The original legislation contained only a few paragraphs on health impacts; SARA had 10 pages. The Agency for Toxic Substances and Disease Registry (ATSDR) was mandated to perform health assessments for each NPL site and for sites proposed for the NPL before the remedial investigation and feasibility study phase was completed by EPA. In other words, ATSDR was responsible for providing a second opinion based on health criteria. To assist in its efforts, ATSDR requested the NRC to review current knowledge of human-health effects by exposure to hazardous-waste sites (NRC, 1991).

In addition, Congress relegated capping and removal and transfer methods to the status of last-resort remedies. It called for on-site use of permanent remedies that reduced the volume, toxicity, or mobility of hazardous substances. At a minimum, on-site treatment must achieve the groundwater quality goals of the Safe Drinking Water Act. These changes reportedly have tended to slow cleanup and to make it more costly (Harris and Wrenn, 1988). Gathering the data required for on-site cleanups resulted in increased expenditure of time and funds. Most important, requiring cleanup to conform to the federal drinking-water standards can be costly. This has become a highly controversial issue.

The January 1991 CERCLA Superfund Inventory System reported the status of more than 33,000 sites. Preliminary assessments and site inspections indicated that 58% of the sites did not pose significant risk. Another 12% awaited site inspection; 18% had completed site inspection; 8% awaited final assessment; and 4% were on the NPL list. EPA reported that remedial actions had been initiated at over 2,000 sites, with cleanup in progress at 400 NPL sites. Yet, EPA has been severely criticized (e.g., Mazmanian and Morell, 1992) because the cleanup had proceeded too slowly. As of February 1994, EPA had deleted from the NPL 57 sites that had been cleaned up and completed remedy construction at another 167 sites, for a total of 224 sites (*Federal Register*, 1994).

In response to the slow rate of cleanup, EPA has proposed a shift to standardized cleanup plans for different types of sites and to shift EPA personnel from nonsite to site cleanup activities. That step, which is likely to take several years to plan and implement, might relieve some of the pressure felt by EPA, but runs the risk that generic programmatic responses might be ill-suited to some sites and communities.

The slow pace of remediation has at least provided the United States with an opportunity to assess the economic implications of different definitions of "clean." Had the pace been faster, more resources might have been committed without an opportunity for reconsideration. For example, Russell et al. (1991) discuss three cleanup scenarios. Each cleanup option assumes equal protection of public health. The goal of the "less stringent" case is to isolate the waste so that people will not be exposed and the uncontaminated environment cannot be contaminated. The "more stringent" case aims to destroy wastes, unless destruction is infeasible or excessively costly. The "current policy" scenario falls between the less and more stringent cases. Russell et al. (1991) estimate a cost of $90 billion to clean up 3,000 NPL sites with the less-stringent criteria. The cost of the current-policy option for the same sites is estimated to be $151 billion, while the estimate for cleanup under the more-stringent criterion is $352 billion. Based upon these estimates, the goal of destroying wastes (more-stringent criterion) costs 3.9 times as does their isolation (less-stringent criterion) and 1.7 times as much as the current-policy option.

Technical Limitations

The Superfund program is unquestionably larger, more complicated, and much more expensive than was envisioned in 1980. In the context of 1980, hazardous-waste sites were repulsive and frightening eyesores of unchecked industrial proliferation that had to be addressed. In the context of the 1990s, the image is unchanged, and perhaps the fear of those sites has even been reinforced. However, the realities of extremely high costs and a program that might take decades to complete have begun to counter the desire to fully clean every contaminated site.

When Superfund was first authorized, there was little experience with cleanup of contaminated soils and groundwater. It was often assumed that existing technologies would meet the need.

However, it soon became apparent that remediation of contaminated sites was much more complex and existing technologies have severe limitations (Hall, 1988; Mackay and Cherry, 1989; NRC, 1990a). As one example frequently cited, pumping and treating groundwater may contain contamination within a site, but not eliminate the contamination. Such treatment is not only costly, but may need to be carried on for decades. It is now recognized that for many sites, there are few if any technical alternatives available for meeting the goal of a permanent remedy. Indeed, while some remedies may address removal of a given contaminant, they might actually have a more adverse effect on human health and the environment than leaving the contaminant alone. Others have discussed the need for legislative changes in Superfund goals that better recognize these important technical limitations.

U.S. DEPARTMENTS OF DEFENSE AND ENERGY PROGRAMS

Chapters 5 and 6 show that DOD's and DOE's models used in priority setting differ in many ways. Nevertheless, this chapter considers the two models together because the contexts of their mandates to manage hazardous-waste sites are similar and because both mandates are markedly different from EPA's Superfund responsibility.

Cleanup Responsibilities

EPA's Superfund responsibility covers every square foot of the United States and its territories, including DOD and DOE installations. DOD and DOE are responsible only for sites they contaminated and for some of the sites their contractors contaminated. That simple distinction in responsibility has had major implications for their development of remediation and priority-setting programs.

DOD and DOE continue to have the mandate to manage their wastes, but changes in the political climate have led to important legal and programmatic changes. DOD recognized off-site contamination from the Rocky Mountain Arsenal (Commerce City, Colorado) in 1974. It began an Installation Restoration Program (IRP) to determine the extent of the problem at the arsenal and at selected other sites (Anderson and Couture, 1984; *Federal Register*, 1989). Executive Order 12316 (*Federal Register*, 1981) and SARA gave DOD the authority to conduct installation reviews and cleanup sites outside the EPA program. However, they also allowed EPA to place DOD and DOE sites on the NPL list, and required the two federal departments to consult with states and tribal governments. Section 120 of CERCLA provided the legal basis for DOE to negotiate three-party interagency agreements with EPA and the states. EPA regional offices were made responsible for the negotiations and oversight of the agreements (Advisory Committee on Nuclear Facility Safety, 1991).

By early 1990, 89 DOD sites were on the NPL (Briefing to committee, April 1991). Eight DOE nuclear weapons complexes are on the NPL (OTA, 1991). Twenty military bases picked for closing are on the NPL (L. Rutsch, EPA, pers. comm., May 1994).

Beginning in 1985, DOE signed agreements with EPA and the states of Colorado, California, Utah, Washington, and New Mexico for many of its biggest facilities, such as Rocky Flats, Hanford, and Lawrence Livermore Laboratory (DOE, 1991a). Those agreements are important for two reasons. They explicitly recognize the importance of EPA and state agencies that are assigned to protect public health and the environment. The agreements are also an implicit priority-setting mechanism—that is, they bind the federal departments to remediating specific sites before all sites have been studied, and they bind the department to cleanup protocols that might have more to do with what the individual state government requires than with the relative risk of the site. That is, there is no guarantee of equity among or even within states in site remediation.

DOD's and DOE's remedial actions and priorities are also influenced by their internal efforts to meet a legacy of other legislative mandates. For example, the Radiation Control Act of 1978 led to the creation of the Uranium Mill Tailings Remedial Action Program (UMTRA) to cover assessments and cleanups at 22 inactive sites and 5,000 adjacent properties. DOE's FUSRAP (Formerly Utilized Sites Remedial Action Program) addresses 31 privately owned sites from the Manhattan nuclear weapons programs, and DOE's SFMP (Surplus Facilities Management Program) focuses on 21 installations associated with civilian nuclear power development (DOE, 1991a). Those programs and other special remediation programs are described in greater detail in Chapters 5 and 6.

In 1991, another legal incentive for cleanup became apparent. DOD's desire to close bases and turn over property was hindered by legal requirements under SARA and directives by DOD that make the government liable for contamination and remedial action on U.S. government sites (P.L. 99-499; DOD, 1987).

Overall, DOD's and DOE's complex, legal mandates constitute an implicit priority system. Resources are allocated to programs based on legal mandates that vary considerably in their demands rather than to sites based on a de novo analysis of risk.

Program Size and Cost: DOD and DOE

DOD's definition of site is similar to EPA's definition of an operable unit. However, there might be multiple sites on an installation that receive the same treatment (e.g., all waste oil lagoons or all drums on concrete pads). In 1988, DOD's IRP included 5,165 sites on 739 installations. Many sites were added over the next few years, in 1991, the IRP reported 17,482 potential sites at 1,855 installations as of September 1990 (DOD, 1991a). Virtually all of these sites have undergone a preliminary analysis. Approximately two-thirds (11,823 of 17,482) have had or were scheduled to have a site inspection. Forty percent have had a remedial investigation and feasibility study scheduled, completed, or performed. Remedial actions had been completed at 296 sites; anohter 1,191 sites had remedial actions under way; and 2,572 sites were scheduled for remediation. In 1991, the General Accounting Office (GAO) reported that, in addition to the 17,482 sites on active installations, DOD had also identified 6,980 sites on land once owned or used and is a potentially responsible party on 185 sites on land where its hazardous waste was disposed of (GAO, 1991). GAO also reported that all the DOD sites potentially requiring remediation have not been identified.

DOE has identified 3,700 sites at 500 facilities (Technical Review Group of DOE, 1991), but the final number of sites might be much higher. Schneider (1991) reported that 17 principal DOE facilities and 50 smaller sites constitute the priority situations.

As in the EPA's Superfund program, actual and projected costs for remediation of DOD sites have escalated rapidly. In 1983 congressional hearings, DOD estimated that 200 waste sites required remediation at a cost of $500 million (U.S. Congress, 1983). In 1985, the General Accounting Office estimated that DOD would spend $5 to 10 billion. However, in 1991, DOD estimated its total cleanup cost to be $24.5 billion (DOD, 1991a). Russell et al. (1991) made "best-guess" estimates for total DOD site cleanup of $18 billion for less-stringent cleanup, $30 billion based upon current policy, and $70 billion for a more stringent scenario. These estimates contrast with those by Shulman (1990), who indicated DOD site cleanup could cost "several hundreds of billions of dollars."

The actual DOD IRP funding was $86 million in 1984, $244 million in 1986, $378 million in 1988, $579 million in 1990, and approximately $1 billion in 1991 (DOD, 1991a). DOD expects IRP funding to increase rapidly and to peak at $2.8 billion per year in 1998. Thereafter, funding is expected to decrease. Seventy-five percent of the money is expected to be spent during the 1990s. Sixty percent of the money would be for site remediation, and 17 percent for operation and maintenance.

Shulman (1990) noted that projected costs for cleanup of DOD's waste-sites might be found unacceptably high, resulting in some zones being set aside as "national sacrifice areas." Alternatively, substantial parts of military installations could provide shelter and habitat for wildlife, especially where units are parts of buffer zones between civilian areas and training areas or operational sites containing unspent munitions. Under a "no-action" option, the sites could remain as sanctuaries for resident mammals and migratory birds and habitats for trees, shrubs and wildflowers. If installations are decommissioned without remediation, thus depriving the area of security, unsuspecting individuals entering the property might inadvertently be injured or exposed to health risks. If cleaned of hazardous materials, such sites could provide recreational land, fisheries, or migratory bird hunting grounds. Once remediated, private and public lands, previously devoted to hazard-associated uses offer opportunity for mixed development (recreational, residential, and light industry) with many options of high value. Thus, there are many options that could lead to conflicting pressures on political, economic and social systems.

Russell et al. (1991) also applied their less-stringent, current-policy, and more-stringent scenarios to DOE remediation costs. Their best-guess estimates were $92-, $240-, and $360-billion for the less-stringent, current-policy, and more-stringent scenarios, respectively. In other words, total DOE costs are estimated to be 5 to 8 times DOD costs. Aggregating DOD and DOE costs gives $110, $270, and $430 billion from 1990 to 2020. Each of the three aggregates is higher than the corresponding estimate for EPA's Superfund program: $110 vs. $90; $270 vs. $151; and $430 vs. $352.

Cleanup costs for many DOE facilities substantially exceed those for the typical Superfund site. For example, a total of $3 to 4 billion has been estimated for the Rocky Flats nuclear weapons plant west of Denver and the Rocky Mountain Arsenal pesticide and nerve gas facilities east of Denver (Schneider, 1991). A total of $4 to 8 billion has been projected to clean up contamination at the Oak Ridge facility, $3 billion for the Savannah River facility, and $30 to 50 billion for the 3,000 sites at the Hanford nuclear reservation (Schneider, 1991). These higher costs reflect the facts that each of these facilities has many sites, and that returning land burdened with both radioactive and chemical wastes to precontamination levels is much costlier than for chemical wastes alone. For example, *Science* (1991) reported a Superfund cleanup of a small radium operation in the basement of a Philadelphia home that produced radioactive elements for cancer treatment. Removing about 1 gram of stray radium required moving 4,000 tons of soil and 1,000 tons of rubble to Utah, at a cost of $11.5 million. Although DOE's site-remediation budget has grown from $1.3 billion in 1989 to $6.2 billion for 1994, the agency does not expect funding to increase as it has in the past (DOE, 1994).

OTHER FEDERAL AND STATE PROGRAMS

Profiles of several other hazardous-waste site-remediation programs, some of which include priority-setting processes, are presented below. These descriptions are not as detailed as the preceding ones for Superfund, DOD, and DOE because less has been written about how priorities are set within these programs. As the accompanying cost projections show, less space in this report does not imply less importance or less need for care in setting priorities.

Leaking Underground Storage Tanks

EPA administers a program to remediate an estimated 300,000 to 400,000 potentially leaking underground storage tanks (OTA, 1989). Based on an average cost of $175,000, Russell et al. (1991) estimate the cost of removing the tanks and remediating surrounding sites to be $67 billion. (A lower limit to that estimate is $32 billion; no upper limit was provided.) EPA, using a technique known as the Analytic Hierarchy Process (Saaty, 1980), plans to focus on water supply impacts in setting priorities for the cleanups (Walsh et al., 1991).

Resource Conservation and Recovery Act

EPA also administers the Resource Conservation and Recovery Act (RCRA) program for active hazardous-waste sites. The best-guess estimate of Russell et al. (1991) is $234 billion (current policy), with a low of $199 billion (less-stringent policy) and a high of $423 billion (more stringent policy). Their best-guess estimate for RCRA is 55% higher than that for the much more highly publicized Superfund program ($151 billion). EPA's ten regional offices use the National Corrective Action Prioritization System (NCAPS) to rank RCRA sites into categories of high-, medium-, and low-potential hazard. Because NCAPS does not produce a numerical score, it requires less data than the HRS does (D. Fagan, EPA, pers. comm., May 1994).

Federal Facilities

In 1990, the U.S. Congressional Budget Office reported potential federal government liabilities for hazardous-waste site remediation at 9,456 facilities. Ninety-three percent (8,805) were DOD and DOE facilities. The remaining 651 were the responsibility of the departments of Agriculture (DOA), Interior (DOI), Transportation (DOT), and other federal departments and agencies.

These organizations are responsible for three types of sites or facilities. DOI's Bureau of Land Management (BLM) must remediate wastes generated at mines, oil and gas exploration sites, and landfills on DOI land. Hazardous wastes resulting from research activities and operations at facilities under the jurisdiction of the U.S. departments of Agriculture, Health and Human Services, Transportation, and Justice; EPA; the National Aeronautics and Space Administration; the Federal Aviation Administration; the Coast Guard; and the General Services Administration constitute a second set of federal liabilities. The third category involves wastes found on lands acquired through foreclosure by the Small Business Administration and the Economic Development Administration.

CBO reported that DOI has the responsibility for 337 facilities with potential remediation requirements; DOT, 101; DOA, 91; and all other federal agencies, 122 facilities. In 1989 and 1990, taken together, approximately $7.5 billion was appropriated by Congress for federal government waste compliance and cleanup (CBO, 1990). Of this, 4 percent ($331 million) was allocated to manage problems at the above sites.

State Programs

States have important responsibilities to manage hazardous-waste sites. If Superfund money is to be used to pay for site remediation, the state must agree to pay 10 percent of the cleanup costs and 100 percent of the operation and maintenance costs (NGA,1994). OTA (1985, 1989) argues that these funding requirements have led states to favor permanent remedies requiring a great deal of up-front funds and as little as possible operation and maintenance costs. The state bias toward permanent remedies, OTA asserts, contributes to higher site cleanup costs because containment and land disposal solutions cost less in the short run but require relatively high operation and maintenance expenditures.

States are also responsible for sites not on the NPL list. OTA (1985, 1989) argues that the majority of states presently lack the funding, technical expertise, and incentive to adequately remediate sites. OTA charges that because state cleanups tend to be less thorough, some will be done poorly and their sites end up on the NPL list. Wright and Cole's (1985) study of risk management of hazardous-waste sites in two states supports that contention. An analysis of four case studies in Texas and Maryland showed state governments as unwilling or unable to provide adequate funds to obtain needed data, analytical capabilities, and technical expertise. On the other hand, in states such as New York, New Jersey, and California, which have strong programs, non-NPL sites might be remediated more rapidly than some on the NPL, with lower litigation and oversight costs, especially if there is a single, non-litigerous PRP. The unwillingness of some states to provide adequate funds, Wright and Cole argue, has led to poorly informed administrators, poorly trained civil servants, confusion about responsibility, and political rather than scientific management of priorities. These limited studies suggest why cleanup of 6,000 to 12,000 sites mandated by state law are estimated to cost anywhere from $3 billion to more than $120 billion (Passell, 1991). The Russell et al. (1991) best-guess estimates of state and private cleanup costs are $18 billion (less-stringent policy), $30 billion (current policy), and $70 billion (more stringent policy) to remediate 30,000 sites at an average cost of $1 million during the period 1990 to 2020.

In 1986, Congress required EPA to involve states in Superfund in "a substantial and meaningful way." Some states are better able than others to work with EPA on NPL sites and on a variety of other hazardous-waste management issues. These differences among states are striking, and not all of them reflect some states having hazardous-waste problems while others do not. According to Schleifstein and O'Byrne (1991) writing in the *Times Picayune* about Louisiana, "we used to say there weren't any hazardous-waste sites in this state, so when Superfund came along New Jersey got all the money." This example illustrates a general finding: states that did not make hazardous-waste remediation a priority, irrespective of the extent of their problem, tended to get relatively few sites on the initial NPL (Greenberg and Anderson, 1984).

The most systematic portrait of state programs is given in the annual analysis by the Environmental Law Institute (ELI). ELI (1991) reports 29 states to be operating cleanup programs supported by enforcement authorities and dedicated funds. Another 12 states have legal authority to conduct cleanups, but lack funding and staffing. ELI (1989) found that South Dakota and Wyoming had no staff for remediation activities, while California, Illinois, Massachusetts, New Jersey, New York, Ohio, Pennsylvania, and Washington each had a staff of more than 100. New Jersey had a staff of over 800 and accounted for more than one-half of the $699 million that states had for remediation. The hazardous-waste site-remediation programs of New Jersey and California are briefly discussed here as examples.

On a per capita basis, some states have made an investment that more than rivals the federal agencies. Ridley (1987) rated New Jersey's hazardous-waste management programs as the most advanced in the United States. New Jersey's cooperative programs with EPA and self-initiated programs illustrate that a state can build a massive program with wide public support.

There were 113 NPL sites in New Jersey, the most of any state, at the time this report was prepared. In addition, 600 of the state's facilities are subject to EPA's RCRA program and approximately 80,000 underground storage tanks storing hazardous substances were regulated by the EPA's Underground Storage Tank Program program and the state's analog. New Jersey also has an Environmental Cleanup Responsibility Act (ECRA, N.J.S.A. 13:1K-6 et seq.), which requires the operator or owner of an industrial facility to clean up any contamination before closing, selling, or legally transferring the site (NJDEPE, 1992). More than 17,000 industrial facilities in New Jersey are regulated by ECRA. In 1993, ECRA was modified and it is unclear at this writing how many facilities will be regulated by the state. New Jersey's Water Pollution Control Act (N.J.S.A. 58:10A-1 et seq.) regulates the discharge of contaminants at over 400 landfills. NJDEPE also regulates the cleanup of illegal spills onto the land and into waterways via its Spill Compensation and Control Act (N.J.S.A. 58:10-12.11a et seq.). Finally, the state has begun a major pollution-prevention initiative. Thus, NJDEPE manages the cleanup and control of hazardous wastes at tens of thousands of sites. In fiscal year 1992, the NJDEPE hazardous-waste program had 750 positions and a budget of $48 million (L. Miller, NJDEPE, pers. comm., February 2, 1992).

Like New Jersey, California has made a major investment in the regulation and management of hazardous waste and hazardous-waste sites. California illustrates a likely trend in funding of state programs that might have implications for setting remediation priorities and the extent of remediation. In the late 1970's, California's original hazardous-waste program was started with approximately 15 people in the Department of Health Services Vector Control Section. By 1985, the effort had grown and was renamed the Toxic Substances Control Division; funding through fees and fines for hazardous-waste site cleanup was supplemented by a $100 million 5-year State Revenue Anticipation Bond. This was the "State Superfund." Funds expended from the State Superfund for hazardous-waste site mitigation and remediation would be collected from responsible parties, along with an assessment of trebled damages. In 1991, the governor of California created the Cal-EPA and within it, the DTSC. Fees and fines, along with monies generated during cost-recovery for State Superfund activities, support continuing cleanup efforts. DTSC now employs over 1,000 persons and has an estimated budget for fiscal years 1992 and 1993 of $110 million. Companies and other responsible parties can conduct their own cleanups, and California's Department of Toxic Substances provides oversight for a fee. For example, fees for oversight of a remedial investigation and feasibility study range from $21,500 for a "small site" to $200,000 for an "extra-large one" (State of California, 1990). Fee-based systems like California's recognize the political reality of public pressure to have sites cleaned up but not to pay directly for the cleanup. There is no way of forecasting the impact of a fee-based system on the type of cleanups that will be chosen and the quality control of those cleanups. There obviously should be some concern that revenue rather than public health and environmental protection will be stressed in a fee-based system.

State Priority Systems

ELI (Environmental Law Institute) (1989, 1991) found 24 states to have ranking systems as an aid to priority-setting. These states do the HRS scoring for EPA and typically choose as state Superfund sites those HRS sites not forwarded to EPA for inclusion on the NPL list. Some states (e.g., Minnesota, New Jersey, Wisconsin) modify the HRS scores.

A few states have their own ranking systems. Montana accumulates information on five factors: (1) contamination of drinking-water supplies; (2) air contamination that threatens public health;

(3) surface water pollution that threatens water used for drinking and recreation; (4) impact on wildlife; and (5) danger of fire or explosion. The state then combines the data without a great deal of quantification to set priorities for cleanups.

Michigan developed its own site assessment model in 1983 (Environmental Response Division, 1990). A later revision scores sites according to six factors: (1) up to 20 points for environmental contamination; (2) up to 5 points for substance mobility; (3) up to 3 points for damage to sensitive environmental resources; (4) up to 4 points for population exposure; (5) 1 point for institutional population; and (6) up to 15 points for the toxicity of waste. Each site can score up to 48 points, but environmental contamination and substance toxicity obviously drive the system.

New York State uses the HRS scores, but has developed separate health and "biothreat" models. The health model emphasizes human exposure and the biothreat model, natural resources damage. Chapter 7 will describe some of the state priority-setting approaches in greater detail.

Setting Priorities

Do DOD, DOE, EPA, and other agencies need scientifically based priority-setting systems? The purpose of such systems is to provide a consistent and scientifically based framework to catalog and compare potential risks to aid in resource allocation, to evaluate progress, and to serve as the basis for communications with affected parties.

Hazardous-waste site-remediation efforts can be started and operated without a scientifically based system. One way this is done is for states to sign legal agreements with the federal agencies or the PRPs that would stipulate remediation based on whatever criteria the parties choose. A second alternative is to allocate cleanup funds directly in proportion to some indirect measure of potential risk. For example, some state-by-state ratio of the number of people dependent upon groundwater to the number of hazardous-waste sites in areas with groundwater might serve as a quick guide for apportioning remedial action funds among state and local areas. Third, the United States has a long history of allocating funds through political processes. Congressional leaders advised by lobbying groups are capable of writing legislation that will assign remediation funds to provide jobs as well as a cleaner environment in their districts. The disadvantage of these nonscientifically based approaches is that places with the largest political influence, because of economic powers or seniority of their elected officials on important political committees are likely to receive a share disproportionate by nonpolitical criteria. In short, without an objective and scientifically based approach for evaluating the relative risk of different sites to human health and the environment, there is no possibility to match risk and urgency with the allocation of remediation funds. One can then only resort to the political process.

Another illustration of how priorities may be set without a scientifically based system is provided in the recommendations of the FFER (Federal Facilities Environmental Restoration) Dialogue Committee (1993). When funding shortfalls result from insufficient appropriations from Congress to meet existing cleanup obligations, the FFER Dialogue Committee recommends a flexible strategy for applying "fair share" principles to allocating funding shortfalls. That is, all federal waste sites subject to outside supervision should share equally in the total amount of the funding shortfall. When a funding shortfall is caused by unanticipated program growth (e.g., due to new circumstances or new data), the FFER Dialogue Committee recommends a greater emphasis on absorbing the shortfall at those sites where it arose and a more general set of principles of fair share relative to shortfalls caused by insufficient appropriations.

EPA, DOD, and DOE have shown different needs for scientifically based priority-setting models. The differences, the committee believes, are manifested in the agencies mathematical models. Chapters 4-6 describe models in some detail. This section suggests how differences among the three agencies, in political pressures and in the numbers and types of sites, are reflected in their models.

EPA funded some engineering studies and emergency remedial actions before Superfund was enacted (Greenberg and Anderson, 1984), but the majority of its remediation funds have been distributed in response to CERCLA and SARA. The idea that sites could be quickly identified and remediated is embedded in CERCLA. EPA required a formal approach that could quickly compare tens of thousands of large sites with small sites, sites in cities with sites in rural areas, and sites in states with strong public support for cleanups with sites in states with little interest in identifying and remediating contaminated sites. Under these circumstances, EPA had neither the time nor funds for 30,000 thorough site reviews and risk assessments. A quick sorting had to be done with a limited amount of data. Indeed, the ATSDR mandate under SARA to provide EPA with a second opinion about sites can be viewed as an effort to alleviate this shortcoming in the CERCLA mandate to EPA.

Until lately, DOD and DOE have not been under as much external political pressure to clean up every site in a few years (FFER Dialogue Committee, 1993). Furthermore, by the time these departments were ready to use a formal mathematical model to set priorities for cleanup of sites, a good deal of their funds were already tied up by legal agreements (Whelan et al., 1987).

DOD's stated goals for setting priorities are to remove imminent health threats, to address the worst sites first, to meet SARA requirements at NPL sites, and to use resources effectively and efficiently (DOD communication with committee, April 10, 1991). DOE's goals call for an allocation system that must be seen to be technically defensible, fair, and not subject to political manipulation (DOE communication with committee, April 10, 1991). In other words, the federal agencies want to remediate sites, use resources efficiently, and protect funds not already committed from being divided by political influence.

DOD's and EPA's models have some obvious resemblances. DOD's Defense Priority Model (DPM) and EPA's HRS are similar in that they are risk-based, include both human health and ecological impacts (though with significantly greater weight on human health); can be used to rank sites; are not amenable to a cost-benefit interpretation; and deliberately omit social and economic impacts from their formal calculations. Like the HRS, the DPM assigns a relative numerical score based on combining and normalizing separate scores for groundwater, surface water, air, and soil pathways. Indeed, both models are also similar insofar as a single contaminant and environmental pathway combination can cause a site to be included on a priority list. The models tend to flag a site by choosing the most serious likely threat rather than precisely assessing risk among the sites.

There are important differences, however. For example, DOD has been able to provide some funding to all sites. Many of these sites are in relatively remote areas, and national security considerations have discouraged any public outcry. DOD has experienced much less pressure than EPA to rapidly make yes or no decisions on every site. Consequently, DOD's model could assume the availability of time and resources to obtain more field data before using its model.

The mathematical modeling approach developed for DOE is different from EPA's and DOD's. In January 1993, DOE estimated that 62 percent of its environmental management activities for fiscal year 1994 are legally driven; 24 percent are other environment, safety, and health activities (required by internal DOE order); and 14 percent are allocated for other desirable program activities (DOE, 1993). T. Cotton (J.K. Research Associates, Arlington, Va., pers. comm. January 1992) and Rezendes (1992) note that results of the DOE modeling approach to setting priorities would not supersede these prior agreements. Cotton also indicated that the numerical results can be used when no agreement is in place and that they can be used to provide a uniform baseline against which all installation requests can be measured. DOD also has agreements with EPA and the states, but the agreements stipulate that DOD can revert to its modeling results when funds are not adequate to cover all agreements (M. Read, DOD, pers. comm., January 1992).

After the committee completed its analyses and was preparing this report, DOD and DOE decided not to use their modeling approach, referred to as the Defense Priority Model and the Environmental Restoration Priority System (ERPS), respectively. Despite those decisions, the committee chose to retain its discussion of these models, because such models serve to broaden the range of priority setting approaches under consideration.

As Chapter 6 shows, the modeling approach DOE was considering (ERPS) included more factors, such as economic impacts and weightings based on public preference. In contrast to the EPA and DOD approaches, ERPS was an optimization-allocation model that attempted to economically optimize the use of resources subject to a series of constraints, most notably legal agreements. This former more complex approach to mathematical setting of priorities was consistent with a goal of maximizing a measure of total return from distributing scarce funds among a limited number of sites. ERPS was not a worst-case-first model—neither DOE's agreements with the EPA regions and the states nor the use of ERPS would necessarily lead to remediating those sites posing the greatest risk to public health and the environment.

PROBLEMS IN THE UNITED STATES

DISCUSSION

Expenditures to clean up all forms of pollution are estimated to be equivalent to about 40% of the U.S. defense budget and just over 2% of the gross domestic product (GDP) (Roberts, 1991). The proportion is expected to climb to 60% of the defense budget and 2.6% to 2.8% of the GDP by the year 2000. Paul Portney, an environmental economist with Resources for the Future, argues for the same public debate on environmental expenditures that occurs for defense spending (Roberts, 1991). There is little doubt that such a public debate has begun, and hazardous-waste remediation is likely to be central to it. Under a current-policy scenario, Russell et al. (1991) estimated plausible lower and upper bounds of $478 billion and $1.046 trillion, respectively, for all hazardous-waste site remediation in the country if it maintains the course on which it has embarked. Their best-guess estimate was $752 billion.

Can and should the United States commit an estimated 400 billion to a trillion dollars over 30 years to remediate hazardous-waste sites? Policies to remediate hazardous-waste sites in the United States have evolved under extraordinary constraints of politics, time, and money. Scientific and engineering analyses have encountered technical problems that defy current consensus regarding applicability of methods of theory. Consequently, scientific and engineering research agendas and recommendations about hazardous-waste remediation often are negotiated and laden with political values. In addition, the growing awareness of the great limitations of available technology to meet the cleanup goals originally envisioned by Congress, and now anticipated by the public, raises serious questions about what actually would be achieved even if this enormous estimated pool of funds were spent for environmental restoration.

The committee believes that the United States is not likely to fund the highest level of estimated costs of remediation discussed in this chapter. Congress and its state equivalents might set aside a relatively constant amount of money to be spent on remediation. In short, EPA, DOD, DOE, and the states are going to have to make difficult choices. One such choice could be that not all sites are going to be cleaned up to their precontamination levels. That is, agencies could decide to clean up some sites to a level suitable for any type of public access, others could be remediated only for limited access, and still others could be declared simply as off limits. Second, the agencies might turn more emphatically to priority-setting approaches and formal models to help choose which sites deserve greater or lesser resources. As discussed in this report, developing a good scientifically-based ranking system is not easy, and much time is required for proper validation. Thus, there is some degree of urgency in starting its development if a sound ranking system is desired.

During the last decade, *in addition to high costs*, five factors have become increasingly evident must be taken into account in prioritizing the cleanup of hazardous-waste sites:

Many priorities are and will be set by legal action. The states as well as other interested parties, have been frustrated for years by the failure of the federal government to move more quickly to clean up sites. DOE and DOD are now subject to many of the same laws as the private sector, and are thus subject to numerous enforcement orders. These orders, usually enforceable in court, not only set forth many procedural steps and regulatory limits, but also contain deadlines for each step and sometimes for ultimate compliance. Although such orders can be renegotiated if all parties agree, it is doubtful whether local citizens would regard a lower priority ranking from a model as sufficient reason for consent.

Lack of availability of remediation alternatives that are technologically and economically feasible. Some projects ranking higher than others might not be good candidates for "ultimate" remediation or even for complete characterization because economically feasible technological solutions have not yet been developed.

Logistics and coordination. Managers must use expensive cleanup crews wisely. On some sites, hundreds of people are deployed along with specialized and complex equipment to manage the sites. Effective planning will be needed to move from one site to another to maximize efficiency and ultimate success.

Political intervention. It is inevitable, as many recognize, that scientifically selected priorities will not automatically be reflected in budgetary priorities. That is the nature of our system and always will be. Sometimes, this effect results in a type of fairness, but often it does not; a good priority-setting system should dampen enthusiasm for this type of administrative or congressional behavior.

Lack of confidence in the existing system. As the body of this report details, there are real considerations that will make any practical system open to legitimate questions. For example, there are obvious reasons to favor a simple, clear scientifically based site ranking model that relies only on known facts, but such a model is found to be to some extent naive, failing to take into account many societal issues and unmeasured technical factors. A bias on the part of some toward cleaning up contaminated sites to a level of zero risk complicates the priority-setting process. Baron et al. (1993) found a preference in many of those surveyed toward cleaning up a waste site completely, even if the total number of lives saved as a result of this strategy would be lower than the number saved if the identical resources had been expended toward cleaning up two different waste sites partially.

The purpose in discussing this list was certainly not to discourage the development and use of a ranking system based on risk information and modeling to aid in making priority-setting decisions. Rather, the purpose of this report is to show something of the context into which such mathematical ranking models must fit for best performance in the overall process of site prioritization and decision-making. Ranking models are not substitutes for political decisions on what to do first, but they can serve as valuable aids to better insure that decisions are not made on political considerations alone. Indeed, they can help decisionmaking by providing factual information of importance in an easy to understand way that will help them make the right choices for the public they serve.

SCOPE OF THE REPORT

The committee examined ranking and priority-setting models developed by EPA, DOD, DOE, and some state governments to help choose sites for remediation from among the tens of thousands of abandoned hazardous-waste sites. It also attempted to understand the larger processes by which these agencies choose sites to remediate and the level of cleanup at each site.

Most of the discussions in this report focus on the site-ranking processes rather than on priority-setting processes, which consider ranked scores along with other factors to arrive at actual remediation priorities. Chapter 2 describes some of the desirable features of a priority-setting system and the analytic models used for site ranking. The chapter provides evaluation criteria applied to methods considered in later chapters. Chapter 3 describes five alternative approaches for evaluations performed as a part of priority setting: risk analysis, environmental impact analysis, structured value-scoring methods, cost-benefit and cost effectiveness approaches, and multiattribute decision making.

Chapter 4 presents the background and history of EPA's Hazard Ranking System (HRS) in the context of priority setting and the legislative mandate and management pressures of the Superfund program. The strengths and drawbacks of the HRS are discussed.

Chapter 5 discusses DOD's development and use of its Defense Priority Model (DPM) to assist in priority setting. The chapter also evaluates the DPM's structure, scoring algorithms, validation, and sensitivity and uncertainty analyses.

Chapter 6 provides an evaluation of DOE's Environmental Restoration Priority System, which can use the results of risk indicator models, such as DOE's Multimedia Environmental pollutant Assessment System (MEPAS). The discussion in that chapter is more descriptive and less analytical than discussions in chapters on the HRS and DPM as the ERPS was not as well documented as the other priority systems.

Chapter 7 examines several of the state ranking models with respect to how they compare with each other and how well they achieve general objectives of a ranking system. The chapter does not describe how the various state ranking systems help to obtain a final priority for cleanup or the policy context of their application.

In Chapter 8, the committee compares and contrasts the procedures used by EPA, DOD, and DOE to make decisions about remediation priorities. A scoring exercise using actual site data and the different agencies' models was performed to help the committee become familiar with the models' input data requirements, operating constraints, and characteristics that contribute to similarities and differences in model outputs.

Chapter 9 discusses the advantages and disadvantages of a unified national process for setting priorities and proposes one such unified process. Chapter 10 presents the committee's general conclusions and recommendations for the overall priority-setting process and the mathematical models that are a part of the overall process. Conclusions and recommendations for the priority setting processes and ranking models of specific agencies are presented in Chapters 4-8.

2

PRIORITY-SETTING PROCESSES

BASIS FOR DEVELOPING A PRIORITY-SETTING APPROACH

The United States faces the challenge of environmental restoration of thousands of contaminated hazardous-waste sites across the nation. Although it is difficult to estimate accurately the total resources required to address this challenge, current projections in terms of the dollars and person-years needed are enormous (see Chapter 1). Given the situation of resources limited by natural catastrophes, federal budget deficits, and other demands, it has become critical that scientifically credible estimates be developed to help choose sites for remediation, determine the extent to which each should be remediated, and set the priority in which remediations occur.

A system to help set priorities for the restoration of hazardous-waste sites could benefit greatly a wide range of individuals and groups involved in environmental restoration, waste management, and public health. The primary goal of such a system is to provide a formal, systematic, consistent, and scientifically based framework to catalog and compare factors to assist in decision making and resource allocation. There are many factors that can affect a priority-setting outcome. These include health, safety, and ecological risks; social and economic values and policies; regulatory requirements; technical considerations, and variation of all these over time. A properly designed priority-setting system would aid decision-makers in (1) designing strategies for minimizing human health and ecological damage; (2) enhancing the sound use of natural resources; (3) promoting the efficient allocation of remediation resources; (4) increasing the efficiency of administrative processes associated with the restoration programs; and (5) strengthening the credibility and acceptance of the priority-setting process.

Most of the discussions in this report focus on *site-ranking processes* rather than on *priority-setting processes*. That distinction is an important one, because most analytic systems developed to date are used only to rank sites according to some numerical score (Halfon and Reggiani, 1986; Halfon, 1989). The ranked scores are then considered along with other factors to arrive at actual remediation priorities (see Chapters 4-7 for specific examples). The site-remediation priorities are, therefore, subject to being different from the numerical rankings. This chapter discusses some of the desirable features of a priority-setting system including the analytic models used in such a system. This chapter also discusses some of the evaluation criteria applied to methods discussed in later chapters.

Although a priority-setting process should focus on individual sites and the feasibility of remediating such sites, the incorporation of such evaluations into a nationwide scoring system and subsequent budget analysis requires careful consideration. For example, even though a ranking based upon the reduction of human-health risk can be used as the basis for national priority setting, the inclusion of issues that pertain to societal impacts might not always have a common denominator nationwide. Different communities or states might place different values on such factors as the loss of wildlife, diminished air quality, or the decline in local real estate values. Thus, the locally affected communities must be involved in the evaluation of sites being considered under any ranking or priority-setting system. An appropriate format for soliciting and explicitly incorporating public input into the priority-setting process is essential.

DESIRABLE FEATURES OF A PRIORITY-SETTING SYSTEM

Overview

A priority-setting system should be designed with an *a priori* knowledge of the purpose and process by which it can affect decision-making. The system should consider the possible solutions in the evaluation process rather than just the severity of environmental impact. For example, a site restoration might have a simple solution, such as the removal of a small amount of contained waste (e.g., in barrels). That might not require a large allocation of resources, and therefore, could be completed in a relatively short time.

Numerous, and often competing, objectives enter into environmental restoration and decision-making. These include the direct and indirect impacts of the hazardous-waste sites on human health and the environment, as well as social and economic effects, at the local level. However, ramifications at the national level must also be considered, especially from the viewpoints of the economic and political impacts of site remediation. The various effects of hazardous waste sites can be categorized as presented in Table 2-1, following the classification of Greenberg and Andersen (1984). Clearly, priority setting is rooted in a multi-objective decision-making process, and thus, it is necessary to have appropriate measures developed for each of the relevant objectives. Although it is tempting to provide a comparative assessment of contaminated sites based on an overall single score that encompasses all factors, such an approach might be unrealistic. A priority-setting method might have to be designed that provides a range of scores that might not be necessarily additive, but might be sufficiently informative to present decision-makers with a clearer view of the problem that they are facing.

Finally, a priority-setting system must have scientific credibility. That is, such a system must be objective and replicable, so as to strengthen its acceptability and effectiveness. The credibility of the system also depends on the accuracy of the data that are available for the sites being considered in the priority-setting process. Uniform requirements for technical data and cost estimates must be established, which will ensure that all sites are evaluated and compared on a consistent basis. At the same time, given that priority setting might be required at many sites in the early stages of investigation—when detailed site information is lacking—the priority-setting process must be flexible enough to handle information at different levels of detail and accuracy, along with the associated uncertainties. The process should (1) allow isolation of those areas of uncertainty that affect site scoring and (2) suggest what additional data should be acquired. That is, readers should be able to tell which information used in the process is of high quality and which information is not. The process should have a mechanism for updating the rankings or priorities as more information becomes available.

TABLE 2-1 Potential Impacts of Hazardous-Waste Sites

Health	Environment	Social	Economic
Immediate death	Ecosystem elimination	Disruption of existing communities	Severe damage of human-made structures
Life-shortening exposure	Elimination of species	Disruption of a few families	Extreme devaluation of property
Acute Illness	Reduction in abundance within species	Disruption of a single household	Reduced appreciation of property values
Severe disability	Reduction in biomass productivity		Loss of productivity of the land
Chronic illness	Loss of use of resource		Local taxpayers pay for cleanup, security, and other site maintenance
Chronic disability	Reduction in use of resource		
Minor or temporary illness			
Emotional illness			

Source: Adapted from Greenberg and Anderson, 1984.

General Issues in Model Development and Application

Because ranking and priority-setting models are designed to be influential components of various environmental restoration programs, it is essential that such models meet high standards. Professionally accepted protocols for proper model development and application should be followed (Gass and Thompson, 1980; Gass and Joel, 1981; Gass, 1983; GAO, 1987).

The following issues concerning model development and application should be considered:

- *Purpose*: A clearly defined and explicitly stated purpose for the model including a defined user population;
- *Credibility and acceptability*: The model's development must include scientific peer review, public participation, and public comment;
- *Appropriate logic and implementation of the model's mathematics*: The equations for evaluating and combining factors must be consistent, scientifically valid, and well chosen for numerical execution;
- *Documentation of the model's development*: Documentation must be provided not only on how to use the model, but also on how the model was developed, i.e., why the model components were chosen over other plausible alternatives;
- *Validation of the model*: The model must have been shown to produce a ranking of site risks or threats reliable enough to fulfill the purpose for which it was designed; and
- *Appropriate sensitivity and uncertainty analyses*: Evaluations must be performed to determine the uncertainties in model scores and the resulting implications for site ranking and setting priorities; appropriate quality control and quality assurance procedures must be incorporated and emphasize quality for input data to which model scores are most sensitive.

PRIORITY-SETTING PROCESSES

Specific Technical Features of a Hazardous-Waste Site-Ranking and Priority-Setting Model

In addition to the general issues and features of the ranking or priority-setting process discussed above, the following technical requirements should be addressed during the development of a computation model used in the process:

• *Model should formally incorporate information regarding uncertainty into its various algorithms when input parameters are unavailable and therefore must be estimated, or when there is lack of confidence in the data.* It is also important that the effects of uncertainties on the final ranking process are clearly identified and reported in a format that is usable by decision-makers.

• *Model should be applicable to all hazardous-waste sites.* The process should be sufficiently flexible to handle all types of hazardous waste sites including, but not limited to, landfills, surface waters and sediments, and contaminated groundwater plumes.

• *Model should allow for dynamic tracking and updating of information.* As new data are obtained about the site and its potential impacts on the surrounding communities and environment, such information should be incorporated into the model; the model should be able to accommodate new information, keep track of the change in priority or rank, and provide a quantitative comparison with prior rankings of the site.

• *Model should discriminate between immediate- and long-term risk to human health and environment.* Long-term risk refers to the potential for harmful effects that might take more than 20 years to be manifest or to site contaminants that pose a risk for long periods (e.g., centuries). A special algorithm may be needed for indicating risk beyond several generations.

• *Model should include cost estimates of remediation alternatives.* These should include considerations of timing related to immediate remediation versus delay in remediation.

• *Model structure should be "transparent."* The various components of the ranking or priority-setting model should be clear, logical, and thoroughly explained; despite possible complexity of the model, the scoring algorithms should be clearly documented and articulated for easy understanding by the users. The model output also should be transparent in terms of providing an overall score and additional information that would allow a person to readily determine why a score is high, and which contaminants, environmental pathways, and receptor populations, etc., are of concern. This additional information is important because different sites could receive high scores for very different reasons.

• *Model should be user-friendly.* The model should be constructed so that nonscientists and individuals who are not computer experts can operate the model; it should be constructed as an interactive system that allows detailed system interrogation and maximum flexibility in generating various scenarios; it should have sufficient on-line help to guide the user through the process of data input and analysis.

• *Model should include appropriate security features to prevent unauthorized changes in site data, model parameters, and model outputs.* Only the system designers and maintenance group should be allowed to make changes to the data bases inherent in the model (e.g., data bases of physicochemical properties or unit risk factors); the system should be protected against tampering and the input of meaningless data (e.g., negative concentration values or values outside certain defined upper and lower limits).

3

CLASSIFICATION OF PRIORITY-SETTING APPROACHES

INTRODUCTION

A variety of priority-setting approaches, such as those employed by DOD, EPA, and DOE, have been developed for specific use to assist in setting priorities for site-remediation efforts or for general use in ranking alternative remedies. The approaches differ considerably according to the single or multiple objectives of priority rankings, the types of data measures used and their degree of uncertainty, and methods for treating intangible—but nevertheless instrumental—factors. Before specific models used in priority setting are reviewed in subsequent chapters, an overview is provided in this chapter of five major approaches that have been applied to evaluation of the possible effects of hazardous-waste site contaminants and to assist in deciding about remediation priorities. The approaches include: risk analysis, environmental impact analysis, structured value-scoring methods, cost-benefit and cost-effectiveness analysis, and multiattribute decision-making. The basic elements of an evaluation of possible effects of environmental contaminants are first identified, then each of the alternative approaches to such an evaluation as an aid to priority setting is discussed briefly. Readers interested in a detailed discussion of these approaches should consult the accompanying citations.

Environmental Evaluation

The process of assessing the potential effects of environmental contaminants, sometimes called environmental evaluation, may be divided into three principal stages (Julien et al., 1992): identification, estimation, and comparison. In the identification stage, the set of environmental elements (e.g., groundwater) and biotic receptors (e.g., humans) that are potentially affected by an activity (e.g., construction of roads or buildings, siting and operation of an industrial plant, or disposal of wastes) are identified, and the types of impacts that could occur are determined. The estimation stage involves estimating the levels of potential impacts including the likelihood, magnitude, and duration of the impacts. In the comparison stage, a synthesis and valuation of the various impacts are made to determine the implications for control or response decisions.

Identification of the potential impacts of an activity is a critical first step in performing an environmental evaluation. Failure to recognize or consider a potential environmental impact has contributed to many of the major environmental problems now facing society, including the legacy of improperly managed hazardous-waste sites. Methods to ensure that the full range of potential impacts is considered for a particular project include the use of map overlays, impact checklists, impact matrices, and cause-effect networks (Julien et al., 1992). These methods are particularly useful for new, large, or one-of-a-kind projects, where previous experience might not be adequate to identify all potential effects. These methods can be used in waste-remediation projects.

In the case of models used to rank hazardous-waste sites, the developers of these models have attempted to be comprehensive in the set of environmental elements and receptors considered and the routes or pathways by which these receptors can be significantly affected. In this sense, the model serves as the organizing structure or checklist for potential site impacts. Through the process of model development, scientific review, and public comment, procedures for site ranking and priority setting might evolve to include a broader spectrum of potentially affected elements. For example, the 1990 HRS revisions, discussed in Chapter 4, added new exposure pathways for human contact with contaminated soils and groundwater-to-surface-water migration, and expanded ecological components to cover a wider range of sensitive environments in the model. The use of models to assist in nationwide priority setting dictates that a common and consistent set of impacts be considered for all sites. Still, the great diversity of local conditions encountered at hazardous-waste sites is such that an ability to consider and incorporate unique and special features of a site is desirable for an evaluation methodology.

The estimation phase in the evaluation of hazardous-waste sites involves the assessment of current or possible future impacts on the biotic receptors and environmental elements at or near the site. That is generally accomplished through the collection of field data and the application of scientific principles to determine or predict (i.e., model) the level or risk of environmental damage. Once the impacts are estimated, a comparison is performed to determine the effect of these impacts on society. As indicated previously, the comprehensive set of impacts possibly due to waste-site contaminants must be considered for the valuation to reflect accurately the potential implications of alternative remediation decisions to society. For example, the hazardous-waste site-ranking models developed by EPA and DOD, discussed in Chapters 4 and 5, respectively do not include explicit consideration of socioeconomic impacts, even though such considerations are critical factors for determining the overall impact of possible remediation decisions. The comparison of impacts inherently involves consideration of values or preferences that may differ for different individuals or stakeholders. It may not be possible to include all variations in a model, but what is important is that the valuation parameters and weights used in the comparison be explicitly stated and separately identified from the scientific parameters in the estimation phase of the environmental evaluation. The methods discussed in the following sections emphasize different approaches to the use of scientific information for impact estimation and comparison for making priority-setting decisions.

RISK ANALYSIS

Risk analysis, or risk assessment, is a qualitative and quantitative process used to evaluate the hazardous properties of a substance and the extent of exposure to them, and to characterize the resulting risk (NRC 1983, 1994a). Risk analysis uses the tools of science, engineering, and statistics to analyze risk-related information and to estimate and evaluate the probability and magnitude of outcomes adverse to humans and other biota (NRC, 1993). Comparative risk analysis can offer a logical framework in which to organize information about complex environmental problems and to assist policy analysts in making resource allocation decisions. It provides an explicit estimate of the likelihood of specific human health or ecological impacts. Risk management is the process of weighing alternatives and selecting a risk-reduction action. Such a process integrates risk-analysis results with engineering data and social, economic, and political concerns to make a decision (NRC, 1983). Major steps in the risk-analysis process, i.e., hazard identification, source characterization, exposure assessment, and risk characterization, are reviewed in the context of hazardous-waste site remediation.

Before performing a formal risk analysis, the site history is evaluated. Research on the past, present, and projected site operations, relations to the surrounding community, and regulatory involvement provides the necessary understanding of the potential nature, magnitude, and degree of contamination. The information collected in this early phase will play an important role in hazard identification, exposure assessment, and risk characterization.

Early in the risk-analysis process, a review of land use at and near the site provides valuable information on the types and frequencies of activities of the surrounding population, and it helps to determine the probability of human exposure by all possible pathways. Identification of the size and characteristics of the populations or individuals most likely to be exposed to contaminants is particularly important in these initial stages of the risk analysis. In addition to demographic information, investigation of community health concerns might provide insight into possible past or current exposures. Examinations of municipal water supplies (recreational, agricultural, and drinking water) for the presence of contaminants can help to determine exposed populations. Moreover, exploring residential and recreational areas can indicate lifestyle factors that lead to exposures or risks to health. Other factors, such as site accessibility and accessibility of the contaminated environmental media (e.g., soil), are examined to make the site-evaluation process more comprehensive and the risk analysis a more reliable means for estimating the effects of a hazard.

Hazard Identification

The identification of potential hazards at a waste site is an iterative process that examines the types and concentrations of contaminants found at hazardous-waste sites. Knowledge of community health concerns, site demographics, and land use provides input to the identification process. Analytical data are evaluated with respect to reliability, accuracy, verifiability, representativeness, and adequacy. Soil, sediment, surface water, and groundwater samples are collected on- and off-site, followed by laboratory testing and direct or statistical data comparisons. Evaluations of sampling data are conducted to determine and rank the potential hazards. Those hazardous agents that exceed legally acceptable levels of concentration are referred to as contaminants of concern. In quantitative risk assessment, the resulting list of contaminants of concern will be investigated further.

Source Characterization

A source term identifies the origin of the contaminant release. A source-release assessment evaluates the likelihood and quantity of contaminant releases from a hazardous-waste site to the surrounding environment. Several types of quantitative techniques may be used alone or in combination during this process: monitoring of environmental contaminants, accident investigation and performance testing, statistical methods, and modeling.

Monitoring focuses on past and current contaminant releases

and involves a regular, ongoing program of sampling in an area near a contaminant source. It is used to detect the type and quantity of the contaminants escaping from the source. Performance testing and investigation under accident conditions provide information on the behavior of systems that might cause a release of toxic substances or materials. This method involves the interpretation of the causes and sequences of events after disruption of a system, as well as the prediction of the system's behavior under a variety of operating or environmental conditions. Statistical methods are used to analyze previously collected data on a risk source, either from monitoring programs or from accident records, to estimate the likelihood of a particular accidental release or hazardous event. Finally, modeling is a formal method employed to estimate key parameters, it requires extensive information about a system's processes, data from monitoring programs, historical event records, or assumptions about probability distributions. Modeling can be used to design improved approaches for other methods. There are several possible models developed to estimate releases, with the choice dependent upon the characteristics of the contaminant source (Cohrssen and Covello, 1989).

Exposure Assessment

Exposure is defined as an event consisting of contact with an environmental contaminant at a boundary between a human and the environment at a specific concentration for a specified interval of time (NRC, 1991). The magnitude of exposure is determined by measuring or estimating the amount of the contaminants available at exchange boundaries (e.g., skin, breathing zone, or gastrointestinal tract) during a specified time period. Exposure assessment is the determination or estimation of the magnitude, frequency, duration, and route (e.g., ingestion) of exposure with regard to both current and future conditions (EPA, 1989a). In order to estimate the level of exposure, the exposure pathways must be identified.

An exposure pathway describes the course a contaminant takes from its source to the exposed individual. A complete exposure pathway links the sources, locations, and types of environmental releases with population locations and activity patterns to determine the significant routes of exposure (*Federal Register*, 1992a). Such an analysis relies, in part, on environmental transport analysis. Environmental transport analysis identifies the mechanism by which released contaminants move through environmental media. There are five major transport pathways through environmental media that are typically considered in estimating health risk: atmosphere, surface soil, groundwater, surface water, and food web contamination.

Risk Characterization

The next step in the risk analysis is to link the potential for exposure to site contaminants with health effects. This part of the risk assessment considers numerous medical, toxicological, demographic, and environmental factors, which determine the potential impact of hazardous substances on human health or, in the case of ecological risk assessment, ecosystem health. This involves quantification or statistical description of the qualitative relationship between a contaminant dose and its adverse effects (response). The human body has complex mechanisms for responding to chemical or biological stimuli; thus, the dose-response phase of risk assessment is highly uncertain and consequently should use all available biological information, including epidemiologic data and animal toxicity studies, to estimate the effects of a given dose of a hazardous substance to a given individual or population. Similar considerations apply for estimates of impacts on plant or animal populations. Dose is the amount of contaminant that is absorbed or deposited in the body of the exposed individual over a specified time.

The risk-characterization phase integrates the previous steps in the risk-assessment process to develop quantitative (e.g., probability distribution) and qualitative estimates of risk. The resulting risk characterization summarizes the estimated human or ecologic impacts, which can be compared to risk-management goals. The expressions of risk developed during the risk-characterization phase are most useful when they reflect uncertainties encountered in the overall risk-analysis process.

Limitations

As with every methodology, risk analysis has limitations. Often, lack of specific data makes it difficult to adequately address critical issues in the risk-assessment process. In these cases, resolution of such issues must be based on professional judgment in addition to quantitative scientific knowledge.

Major criticisms of the risk-analysis process include the following: (1) risk assessors might manipulate the risk-analysis process to produce a desired conclusion, (2) many important factors cannot be incorporated adequately into a risk assessment, and (3) that risk analysis does not possess a sufficient level of precision to be used in priority setting. That is, there is too much uncertainty in the results.

First, there is concern that the risk assessor might manipulate the process to produce a personally desirable conclusion. The value of a risk analysis is that the process requests explicit statements of the steps and assumptions used in deriving the risk estimate. Typically, the results of risk assessments are subject to critical review by other scientists and managers, thus offering an opportunity to reveal structural or procedural errors, manipulation, or arbitrariness during the review process. Still, many contend that, in certain cases, this review process has not been sufficient, and that risk assessments have been skewed for purposes of supporting predetermined conclusions. To avoid this, care is needed to ensure that risk-assessment studies are conducted in an open manner with active public participation.

Second, there is concern that many factors cannot be incorporated adequately into a risk assessment. Within the domain of human health, for example, there may be a number of concerns (e.g., birth defects and neurological effects) in addition to cancer. Adequate techniques and data may not be available to assess the risk of cancer and noncancer health effects that are of concern.

Finally, risk assessment is sometimes criticized for not being precise enough to be used in environmental decision-making and setting priorities. Risk analysis is indeed a process that involves much uncertainty, but the existence of uncertainty in and of itself should not disqualify its use to aid in priority setting. For example, the prediction of health and environmental effects rests upon extrapolation of an assumed relation between a dose and a particular type of response. By improving mathematical models used to produce risk estimates and expanding risk-assessment data, uncertainty in risk analysis can be reduced. To help users understand better the results of a risk estimate, risk analysts must indicate the strength of support for the estimate. Therefore, the statistical descriptions of risk produced by risk analysis should include measures of variance or confidence levels to indicate the strength of support for each risk estimate.

In the context of government decision making, risk assessment is followed by risk-management activities. People perceive risks differently depending on the nature of the risk, individual experiences, trust in authority, and efforts to communicate risk (NRC,

Classification of Approaches

1989). Individuals, organizations (e.g., news media, interest groups), and governments often make decisions based on perceptions of risk. The tools of social, economic, and political sciences are employed to help the public better understand risk information through effective communication. Risk-assessment techniques provide the risk manager a means of organizing relevant information and estimating adverse health effects or environmental impacts (NRC, 1994b). In the final analysis, risk assessment—however imperfect—represents a best attempt to set forth what is known in order to aid a decision in the face of uncertainty.

Environmental Impact Analysis

The National Environmental Policy Act (NEPA) of 1970 (P.L. 91-190) was intended to raise awareness of the environmental consequences of new projects. It mandated environmental impact analyses of substantial new industrial, commercial, and public works projects. The environmental impact statement (EIS) requirement applies to federal agencies. State and local governments have also made the EIS a requirement for many government and private projects.

The EIS was a formal tool for balancing economic growth considerations against the effects of pollution on air, land, and water as well as other external effects. Federal agencies were obligated to analyze the impacts of their projects, to consider alternatives, and to take steps to ameliorate serious adverse impacts (Odell, 1976). The U.S. courts have played a major role in determining the scope of the EIS requirement by defining "significant action," "major action," and the parties who may sue for noncompliance.

Based on its goals, environmental impact analysis (EIA) would appear to have little in common with the site-ranking models designed for DOD, DOE, and EPA. The EIA process is intended to be preventive. NEPA and the succeeding legislation require a balance between development goals and environmental protection on a project-by-project basis. The initiating agency (private or public) is required to demonstrate that its development design will not need to be canceled or modified because of environmental considerations. However, critics of the EIA process charge that analyses support development projects by deliberately understating or ignoring serious environmental impacts and are rarely open to alternative designs that might alleviate problems.

In contrast, DOD, DOE, and EPA models are intended to be objective analyses to aid in making priority-setting decisions. Advocacy or opposition on the part of federal, state, and local governments; corporate officials; citizens groups; and other interested parties theoretically do not enter the scoring process for sites. However, advocacy pressures can influence the site-scoring and priority-setting processes as well.

Nevertheless, there are two important similarities between the models to assist in priority setting for hazardous-waste site cleanup and the EIA process. They both attempt to cover a broad range of effects on water, air, and land quality, although public health dominates the models used in priority setting, and environmental protection dominates the EIA requirement. Because of their need to compare different effects, both approaches impose a simplifying and integrating quantitative structure on disparate information. For example, the Battelle EIA approach for a proposed water project considers categories of information on the physical and chemical impacts on the body of water and the ecological, aesthetic, and social effects on the surrounding area (Dee et al., 1973). Scales and weights are assigned to each of these impacts and, like their equivalents in the EPA, DOD, and DOE models, these scales and weights sacrifice information about some variables and impose a quantitative structure on others to arrive at an overall score. The ranking models and EIA process can be challenged for blending together strong and weak data sets, for oversimplifying or ignoring theory, and for being difficult to validate because of their metrics. Furthermore, many of these early EIA processes were the scientific forerunners of EPA's Hazard Ranking System (HRS) and DOD's Defense Priority Model (DPM).

The ranking models and EIA processes have another similarity—a political function. The EIA obligated government agencies, including the U.S. Department of Transportation, the U.S. Department of Agriculture, and the Army Corps of Engineers to consider environmental impacts along with their historic missions. That consideration was required to be explicit and open. Likewise, the site-ranking models have the potential of making the process of setting priorities available to public review, scrutiny, and comment.

The EIA process has gone through numerous revisions. Each revision addressed differences between those who wanted to make it more inclusive and precise (more variables, better data, and more rigorous scientific standards) and those who wanted to simplify the process because its already high cost and complexity seemed to obfuscate rather than clarify impacts. The DOD, DOE, and EPA approaches to ranking sites could benefit from changes made in the EIA process during the last two decades, particularly from the success of the EIA in balancing the desire for greater comprehensiveness and the need for simplicity.

With respect to their user friendliness, some of the EIA approaches are well documented and easy to follow. For example, the EIA approaches presented by a Canter and Hill (1979) and Inhaber (1976) are particularly clear. The authors described the assumptions and strengths and weaknesses of each parameter and index used in their models. That level of clarity is lacking in the site-ranking models addressed in this study.

STRUCTURED VALUE-SCORING METHODS

The limited availability of suitable theory, algorithms, and data can often rule out the application of rigorous scientific estimates and models at the stage when environmental decisions, such as site assessment and priority setting, must be made. For these cases, more qualitative or heuristic approaches have been developed to act as surrogates for formal scientific risk assessment. These techniques constitute what has been referred to as a structured-value approach (EPA, 1988; Carpenter, 1990), and it is used in HRS and DPM for site scoring.

A structured-value approach incorporates in a mathematical framework the major input factors that determine impacts and risk, but it does so in a heuristic manner. Field data and qualitative judgment are used to assign scores for different levels of the input factors, and these scores are combined mathematically to obtain an overall score for a particular potential impact. The scoring categories often reflect scientific knowledge and expertise on indicators such as pollutant release, mobility, exposure, and impact, but they are not rigorously comparable to, or testable against quantitative measures of these indicators, which are used in formal risk analysis.

Risk-analysis models multiply factors obtained from environmental transport and dose-response algorithms to provide an estimate of risk (e.g., Crouch and Wilson, 1981; Crouch et al., 1983; Pushkin, 1992). In contrast, structured-value models often involve additive or weighted sum calculations, although a variety of mathematical functions can be used, subject to the judgment of the model developers. The developers may in fact have had in mind the multiplicative model for risk when they selected the algorithms, but chose an additive model to correspond to a logarithmic scale for the input factors and the resulting risk estimate. Factor scores are generally combined in such a way as to yield a scaled result for each of the impact measures (e.g., between 0 and 1, or 0 and 100) to allow subsequent aggregation of the impacts into a single score.

The major disadvantage of the structured-value model is that rigorous scientific validation and testing are not possible. The heuristic, judgmental nature of the scoring procedure and the dimensionless, scaled nature of the model output preclude comparison with observed data in any absolute sense, and even comparison of risk indicators is difficult, except for the ordinal result that a higher score should be worse than a lower score. Because the scores only provide risk rankings in an ordinal sense, they cannot be used to compare the benefits of alternative environmental decisions, such as the implementation of different remediation actions at different sites. As such, reductions in the structured-value score that are observed or projected as a result of remediation activities cannot be used rigorously to quantify the benefits of these activities.

The structured-value approach is also used in the comparison stage of an environmental evaluation. In the comparison stage, the estimated effects are combined to obtain an aggregate measure of potential impact due to contaminants at a hazardous-waste site. This step requires the assignment of value or importance factors for the various impacts, even in the case where scientifically rigorous methods are used to estimate these impacts. Linear weighting or various other algorithms can be used to determine an aggregate measure of potential impact or importance.

The use of scoring methods to aggregate impacts in the comparison stage has a long, although controversial, history in the domain of multiattribute utility theory discussed in the next section. Scoring of various environmental measures is also used in a number of the formal procedures for Environmental Impact Assessment (Inhaber, 1976; Canter and Hill, 1979; Thompson, 1990). The procedure, when applied rigorously and openly, can provide useful guidance for multiattribute decision problems.

In weighting different impacts for aggregate evaluation, there is no definitive approach, only different views. The algorithms and weighting factors used in structured-value models typically represent the consensus values and preferences of those who have developed, reviewed, and approved the model. Those might or might not appropriately represent the views of all interested groups and affected parties or of society as a whole. Although comparison methods have been developed to document the distribution of impacts among stakeholders with different values and goals (Lichfield et al., 1975; Davos, 1977; McAllister, 1980), these methods are particularly difficult to apply across multiple projects with a wide diversity of interested parties.

For the site-ranking models considered in this report, the value weights are often hidden in the algorithms, and thus it is difficult to separate the factors that represent scientific procedures from those that imply value judgments. Such separation is essential for an effective understanding, critique, and use of the models (Hyman and Stifel, 1988). In addition, decision-makers must have access to the reasoning process used in the development of the value weights (Westman, 1985). The output from a site-ranking model should thus provide information in addition to the overall score itself, so that one can understand why a high or low score was obtained. The additional information could include individual environmental pathway scores, whether site contaminants pose acute or chronic risks, and how the model's value-weights affect the overall score.

Multiattribute Approaches

Multiattribute approaches involve systematic and documented techniques for aggregating subscores (or developing composite scores) that involve subjective values and scientific judgments.

CLASSIFICATION OF APPROACHES

Each of the two techniques mentioned next has an explicitly theoretical basis and is best applied with the guidance of an experienced professional. Each could be applied not only to the final aggregation of site scores but also at a point in the process that develops subscores.

The Multi-Attribute Utility Technique (MAUT) has a strong theoretical basis and has been widely used. Keeney and Raiffa (1976) present a well-regarded treatise on the technique; Keeney (1977) has conveyed the gist of the technique in the context of an application involving environmental effects of energy generation. Edwards and Newman (1982) and Hammer et al. (1988) provide additional information.

The Analytic Hierarchy Process (AHP) handles weighting through analysis of a matrix, the entries of which estimate the relative importance of the attributes associated with pairs of subscores. There is explanatory theory for this weighting, and software to support the necessary calculations. The AHP, more controversial than the MAUT, has been gaining acceptance. Saaty (1980) wrote the classical treatise on the method; and Golden et al. (1989) produced a volume that contains a number of case studies (Paper 3 referenced by Golden et al. lists over 150 application papers). Criticisms of the method (e.g., relative ranking of alternatives can be upset by the addition of another alternative) are highlighted by Dyer (1990), with counterarguments by Saaty (1990) and Harker and Vargas (1990).

EPA's HRS and DOD's DPM use weights for the separate elements included in the scoring process, but the methods for aggregating subscores are not adequately justified by an analytical explanation. DOE's Environmental Restoration Priority System has an explicit and formal multiattribute utility basis combining estimates of human health, environmental, socioeconomic, and regulatory benefits of remediation.

81

Cost-Benefit and Cost-Effectiveness Approaches

Rational public decision-making implies a process for determining appropriate action by utilizing scarce resources in such a way as to maximize the attainment of given objectives. Government agencies have the prime responsibility for carrying out their legislative mandates within each program by selecting the activities that best fulfill their basic mission, have highest priority, and can be ranked at lower levels (or rejected) because the activities contribute little, not at all, or negatively. As discussed here, cost-benefit and cost-effectiveness approaches for environmental evaluation have been adapted from private-to-public-sector use for assisting policymakers to achieve well-defined goals when resource constraints require the ranking of alternative courses of action.

Cost-benefit analysis is a technique for evaluating alternative courses of action when inputs (costs) and outputs (benefits) can be compared based on the same metrics (e.g., monetary values) (Prest and Turvey, 1969; Lave and Gruenspecht, 1991; Krupnick and Portney, 1991). Risk-benefit analysis is a similar approach, but different to the extent that the costs of hazard reduction (and often the benefits as well) are subject to much uncertainty and are expressed in terms of a distribution of possible outcomes with associated "expected" levels.

The cost-effectiveness methodology is used when inputs can be assessed in market values but outputs cannot be evaluated in dollar terms. Thus, costs and benefits of alternative courses of action can be compared with each other within but not across program areas.

Cost-benefit and cost-effectiveness approaches share three basic ways of structuring priorities:

- select the activities in order of increasing cost (rank activities that achieve a specified level of output with the least cost);

CLASSIFICATION OF APPROACHES

- select the activities in decreasing order of benefit or effectiveness within a given budget constraint (maximize benefits subject to a specific level of cost);
- allow activities and their decision parameters to vary, evaluate the resulting variations in costs and benefits, and then rank activities according to the *ratio* or the *difference* (whichever is more appropriate) between benefits and costs.

A variant of cost-benefit and cost-effectiveness approaches explicitly includes tabulation of the incidence and distributional (e.g., ethnicity, gender, age, spatial) effects of costs and benefits among affected groups. The tabulation makes it possible to track which groups are likely to receive net benefits from each of the proposed activities and which are likely to be harmed. This incidence approach allows policy-makers to include distributional or equity criteria into a ranking scheme (Hill, 1968). The incidence matrix can include costs and benefits that are quantifiable but cannot be expressed in monetary terms. It can also include costs and benefits that can be identified as nominal inputs and outputs but are more intangible and not measured in a common metric.

Because scores produced by structured-value models are not necessarily proportional to any measure of utility, it is difficult to apply cost-benefit and cost-effectiveness approaches to such scores. Therefore, neither EPA's HRS nor DOD's DPM provides explicit consideration of the costs of remedial actions. They are intended solely to rank sites such that those sites are identified where human health or ecological risk could justify remedial activity. In contrast, the DOE approach identifies both the benefits and the costs of alternative remedial actions for guidance in allocating resources.

The economic-related approaches described in this section are not without limitations. One is difficulty of obtaining appropriate information on all the costs and benefits. Typically, costs are relatively easy to quantify, but economic benefits are more difficult to measure. The ability of these approaches to predict future economic outcomes is difficult because resource values change.

4

EPA's Priority Setting

The U.S. Environmental Protection Agency (EPA) has primary responsibility for environmental management and regulation in the United States, and with it, the authority to identify the most serious abandoned hazardous-waste sites for attention under the federal Superfund program. As part of this authority, EPA must determine criteria for inclusion on the National Priorities List (NPL) for Superfund sites and the pace at which sites continue along the administrative path from identification to remedial action and closure. The principal priority-setting step occurs when a site, following preliminary assessment (PA) and site inspection (SI), is scored using the Hazard Ranking System (HRS) model. The score (ranging from 0 to 100) determines whether the proposed site is included on the NPL and remains under the continued auspices of the federal Superfund program. Other scoring and ranking systems are used by EPA in other phases of the Superfund program, although as shown later, they are considerably less formal and rigorous.

In this chapter, the background and history of the HRS are presented, and the model's approach and structure are characterized and evaluated. The evolution of the HRS is traced in the context of the legislative mandate and management pressures that have guided and constrained EPA's administration of the Superfund program. The strengths and drawbacks of the HRS are discussed, with particular focus on changes that occurred with implementation of the revised HRS in December 1990.

Background and History

With the realization of the magnitude and potential impact of hazardous-waste contamination that occurred following the Love Canal incident in 1978, Congress passed the Comprehensive Environmental Response, Compensation, and Liability Act (CERCLA) of 1980. That law granted EPA the authority to respond to current or potential releases of hazardous waste that could threaten "public health or welfare or the environment." It established the principle of strict, joint and several liability whereby all potentially responsible parties (PRPs) identified at a site are liable for the costs of addressing and removing the hazardous threat. A multi-billion dollar fund was established through taxes on petroleum and chemical feedstocks to pay the costs of response action and remediation in cases where viable PRPs were not present or in cases where immediate federal action was deemed necessary. This Superfund, administered by EPA, has since provided the name by which the entire CERCLA process and the sites themselves have become known.

The initial CERCLA legislation was debated and passed under highly charged conditions, and many of the involved parties, including EPA, were primarily concerned with establishing their position at the forefront of this new and powerful tool for harnessing public concern and anger over environmental contamination (Landy et al., 1990). Many practical issues of implementation, such as methods for setting site priorities, were largely ignored in the development of the legislation. CERCLA did require EPA to establish "criteria for determining priorities among releases or threatened releases [of hazardous substances] through the United States for the purpose of taking remedial action." Furthermore, the "criteria and priorities . . . shall be based upon the relative risk or danger to public health or *welfare* or the environment" (emphasis added) (CERCLA, 1980, Section 105(8)(A)). These criteria were to take into account the following considerations as much as possible:

- the population at risk;
- hazard potential of substances at the facilities;
- potential for contamination of drinking water supplies;
- potential for direct human contact; and
- potential for destruction of sensitive ecosystems

As highlighted above, the initial criteria and priorities were to consider public health, the environment, and public welfare. The HRS, however, is designed to focus solely upon human health and the environment, with socioeconomic impacts considered only in an indirect manner.

To determine which candidate sites would be included on the NPL, EPA contracted for the development of the original HRS model. The HRS model was developed by the MITRE Corporation to meet EPA's need for a multimedia environmental assessment model (Chang et al., 1981). At that time, multimedia assessment procedures were unavailable, and although pollutant transport and fate models had been developed for some of the individual pathways considered, those models were not connectable or comprehensive. Furthermore, methodologies for environmental and health risk assessment were just beginning to be developed. The multimedia risk approach of the HRS model was thus very innovative for its time. Following scientific review and public comment, formal adoption of the HRS occurred with passage of the National Oil and Hazardous Substances Pollution Contingency Plan (40 CFR 300), which indicated that the original HRS would be "used to assess the relative threat associated with actual or potential releases of hazardous substances at sites" (Appendix A, 40 CFR 300).

Through the 1980s, dissatisfaction with the HRS, motivated in part by a desire to provide a more accurate representation of relative risks, particularly at large coal and other mining facilities, led to the push for modifications of the HRS. The requirement for modifications was included in the Superfund Amendments and Reauthorization Act (SARA) of 1986 which instructed EPA to amend the HRS to ensure, "to the maximum extent feasible, that the hazard ranking system accurately assesses the relative degree of risk to human health and the environment" (SARA, 1986, Section 105(C)(1)). It is noted in SARA that, given the need for expeditious site identification, the revised HRS is not required to be equivalent to a detailed risk assessment, but rather should be as accurate as possible using the screening level information usually available at the preliminary assessment (PA) and site inspection (SI) phases of the Superfund process. Further requirements of the mandated revisions included the need to consider potential and observed air contamination; effects through the human food chain; and better risk assessments for large-volume wastes, including the quantity, toxicity, and concentration of wastes and their potential for release to the environment. The target date given in 1986 for SARA-mandated revisions was October 1988; however, final promulgation of the revised HRS did not occur until December 1990 (*Federal Register*, 1990).

EPA's Priority Setting

Role of the HRS in the Superfund Program

The primary function of the HRS is to serve as the screening mechanism for determining which candidate sites are included on the Superfund NPL. The major steps in this process are summarized in Figure 4-1. A site where hazardous-waste problems are known or suspected is first placed on the CERCLA Information System (CERCLIS), which is the master list of hazardous-waste sites in the United States. A PA and SI are conducted to provide a screening evaluation of the site and to gather the necessary information for scoring the site with the HRS. The sites are scored under the auspices of a regional EPA office by designated contractors or state agencies and submitted to that office for review. The site is proposed for placement on the NPL if the final HRS score is greater than or equal to 28.50; if the score is below 28.50, the site is designated as "no further remediation action planned (NFRAP)" under the federal Superfund program. The selection of the 28.50 cutoff score was initially made in 1982; it was chosen to meet the CERCLA mandate (CERCLA, Section 105(8)(B)) that at least 400 of the approximately 700 CERCLIS sites first scored at that time would be included on the NPL. The cutoff number thus had no apparent significance in terms of an absolute level of environmental or human health risk. Sites proposed for the NPL as a result of their HRS score undergo a period of public comment, after which the final decision for inclusion on the NPL is made by EPA. Through February 1991, only 79 sites had been proposed but rejected for inclusion on the NPL, in most cases because their revised HRS score was below 28.50 or because the site was reclassified as a Resource Conservation and Recovery Act (RCRA) facility (EPA, 1991a). Other mechanisms are also available for placement of a site on the NPL. States are each allowed to nominate one high priority site irrespective of its HRS score. As of 1992,

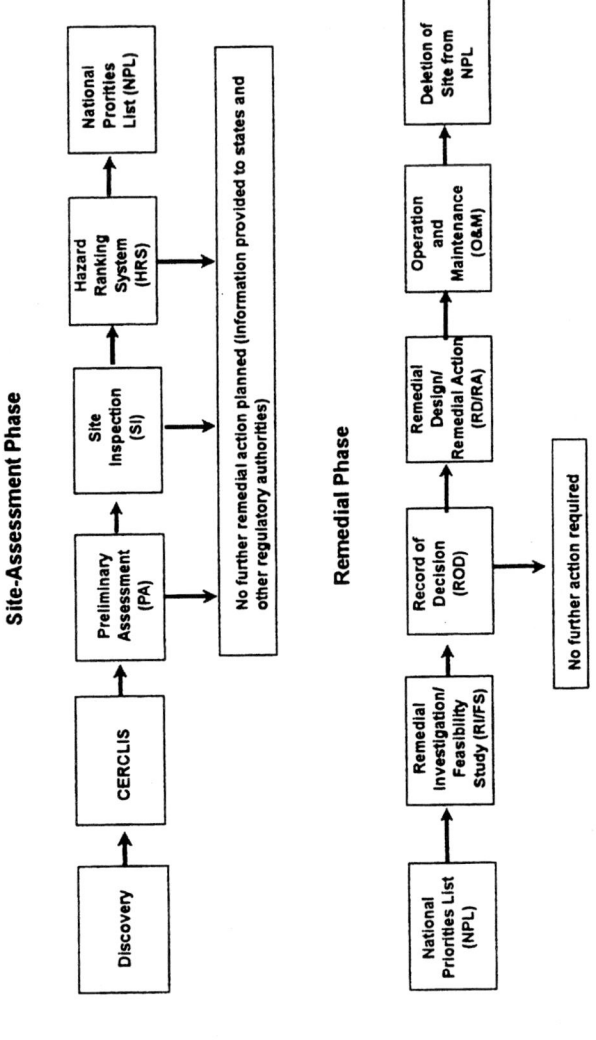

FIGURE 4-1 The Superfund process and the role of the Hazard Ranking System

approximately two-thirds of the states had proposed sites in this manner (EPA, 1991b). In addition, a site can be placed on the NPL as a result of a health advisory from the Agency for Toxic Substances Disease Registry (ATSDR). Five sites have been proposed through this mechanism. With more than 1,200 sites now on the NPL, the HRS score has been the principal mechanism for determining whether candidate sites are nominated and included. This score thus serves a critical role in determining the priority and level of attention that a site will receive in the EPA Superfund program.

The steps shown in Figure 4-1 provide an idealized and highly simplified representation of the Superfund site-selection process and the role that the HRS plays in it. In actual practice, the process is more involved, as summarized in Figure 4-2. As shown near the bottom of the diagram, the HRS scoring must undergo review by EPA headquarters and a quality-assurance (QA) review by a contractor before the decision for nomination to the NPL. Furthermore, simplified screening versions of the HRS have evolved to allow sites to be prescored following the preliminary assessment (PA). The PA method is based upon the full HRS, but uses conservative default values for factors that are still unknown at the conclusion of the preliminary assessment phase of the analysis (EPA, 1991c). The PA method is designed to result in a score that is at least as high as the subsequent HRS score, and can therefore serve as a screening mechanism. The intent is to avoid, where possible, the expenditure of time and resources on sites where the potential for eventual inclusion on the NPL is low or nonexistent. The development of screening steps designed to eliminate false positives along the site-selection process has been largely motivated by the difficulties EPA has encountered in attempting to process the large number of CERCLIS sites under consideration for the NPL. As discussed later in this chapter, management considerations of this type, rather than environmental evaluation, have often been the driving factors in the evolution

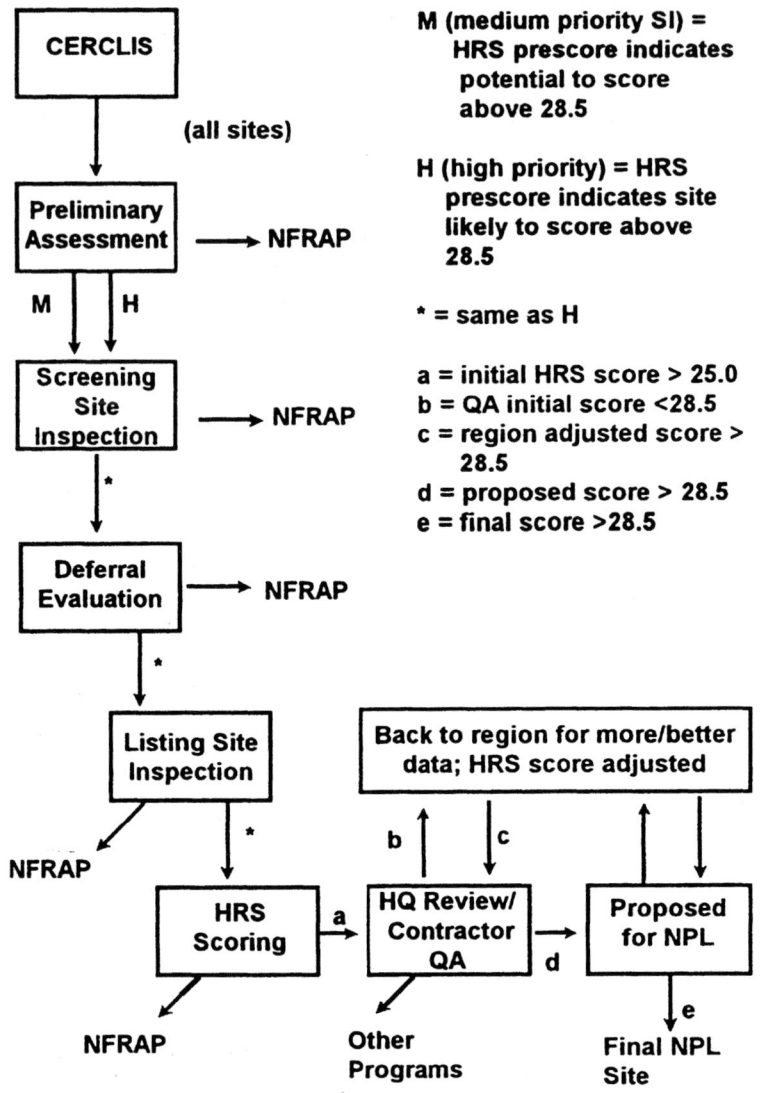

FIGURE 4-2 More detailed steps in the HRS scoring process and National Priorities List development. Source: OTA, 1989.

of the Superfund site-selection process and have subsequently affected the role of the HRS in this process.

In addition to its formal role in the NPL selection process, the HRS model has been used by others for various purposes. Some states use the HRS score to set priorities for sites under their jurisdiction that do not qualify for the NPL. Many states implemented their own site-scoring systems, which were often similar to the original HRS model. The HRS score is used by EPA regional offices as a starting point in their subsequent remedial investigation and feasibility study (RI/FS) priority process, discussed later in this chapter. A positive correlation between the HRS score and the pace of subsequent Superfund actions has been found by Hird (1990). Finally, the HRS score has been proposed as a general mechanism for quantifying risks from hazardous-waste sites and measuring the risk reduction achieved in subsequent remediation (Wilson, 1991; Butler and Jones, 1992).

Model Structure and Components

The HRS is a structured-value model in which various characteristics of the site, wastes, and surrounding environment are combined through use of a numerical algorithm to compute an overall score. As part of the calculations, separate scores are computed for each of four exposure pathways:

- groundwater migration pathway (S_{gw});
- surface water migration pathway (S_{sw});
- soil exposure pathway (S_s); and
- air migration pathway (S_a).

The overall score is determined as the root mean square average of the four pathway scores:

$$S = \left[\frac{S^2_{gw} + S^2_{sw} + S^2_s + S^2_a}{4} \right]^{1/2}$$

That score and each of the individual pathway scores range from 0 to 100, with higher scores reflecting higher degrees of threat. A schematic summary of the major components and calculations of the revised HRS model is presented in Figure 4-3. The algorithm is structured to include the effect of three factor categories:

- likelihood of release or exposure;
- characteristics of the wastes present at the site; and
- characteristics of the target population or environment.

The score for each pathway is calculated as the product of its three factor category scores. The likelihood of release is determined based on the presence of an observed or potential release. Observed releases are verified with site monitoring data. The potential for release depends on pathway characteristics that either restrict or promote transport at or near the site.

The waste characteristics are chemical-specific and are intended to represent the properties of the chemical that indicate the likelihood of exposure and potential health hazard. The waste characteristics considered across all pathways include the toxicity, persistence or mobility, and hazardous-waste quantity. The bioaccumulation potential is considered in the surface water migration pathway for human food chain and environmental impacts.

The environmental and human health targets considered in the HRS vary across pathways. The groundwater migration pathway includes water supply wells, groundwater resources, and wellhead protection areas. The surface water migration pathway considers drinking water intakes, human food chain impacts, and sensitive environments. Calculations for the soil exposure pathway include potential health impacts to residents and workers on-site and

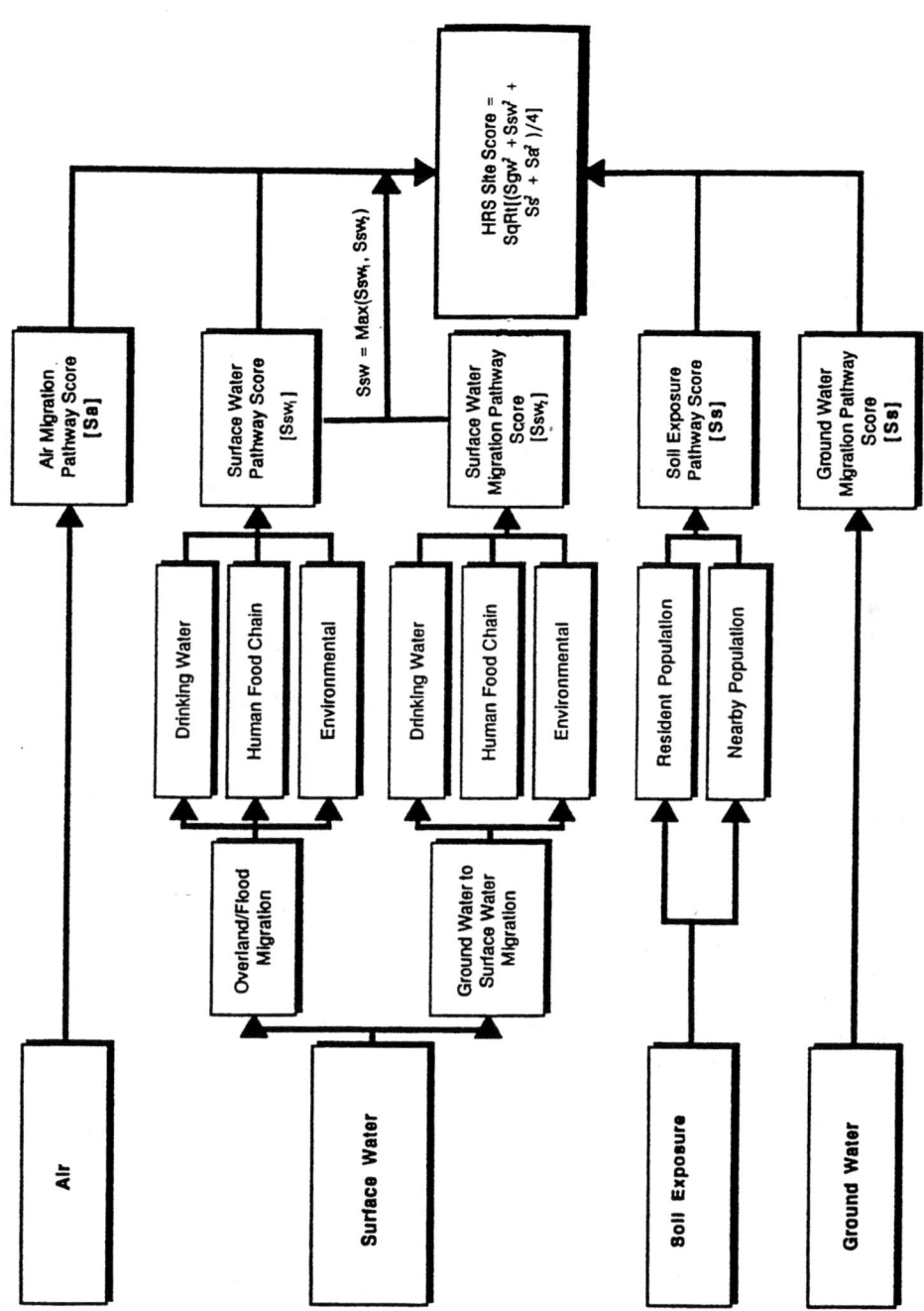

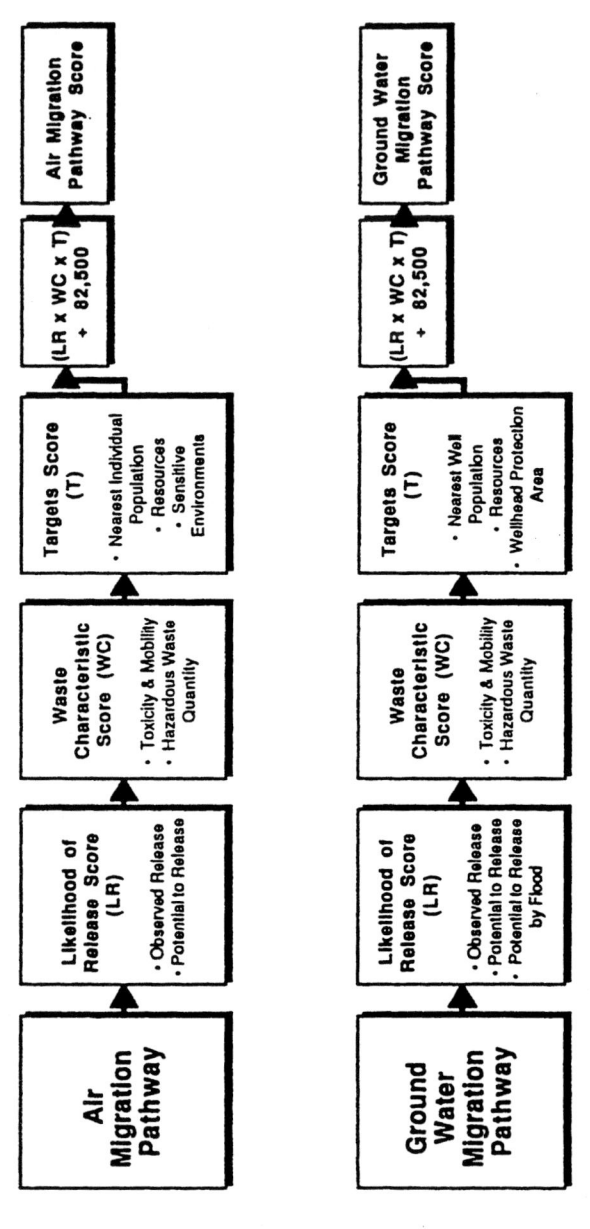

FIGURE 4-3B Individual pathway calculations of the HRS.

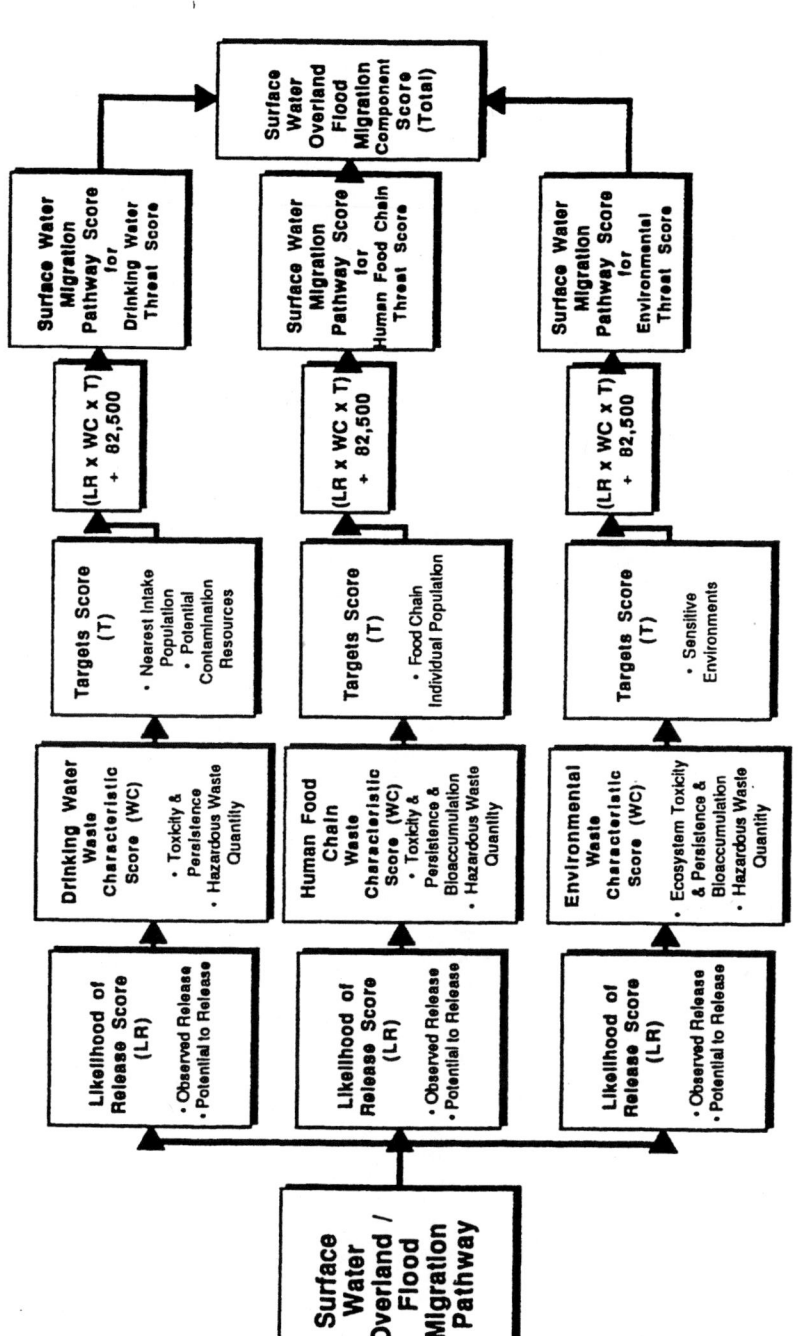

FIGURE 4-3B (CONTINUED)

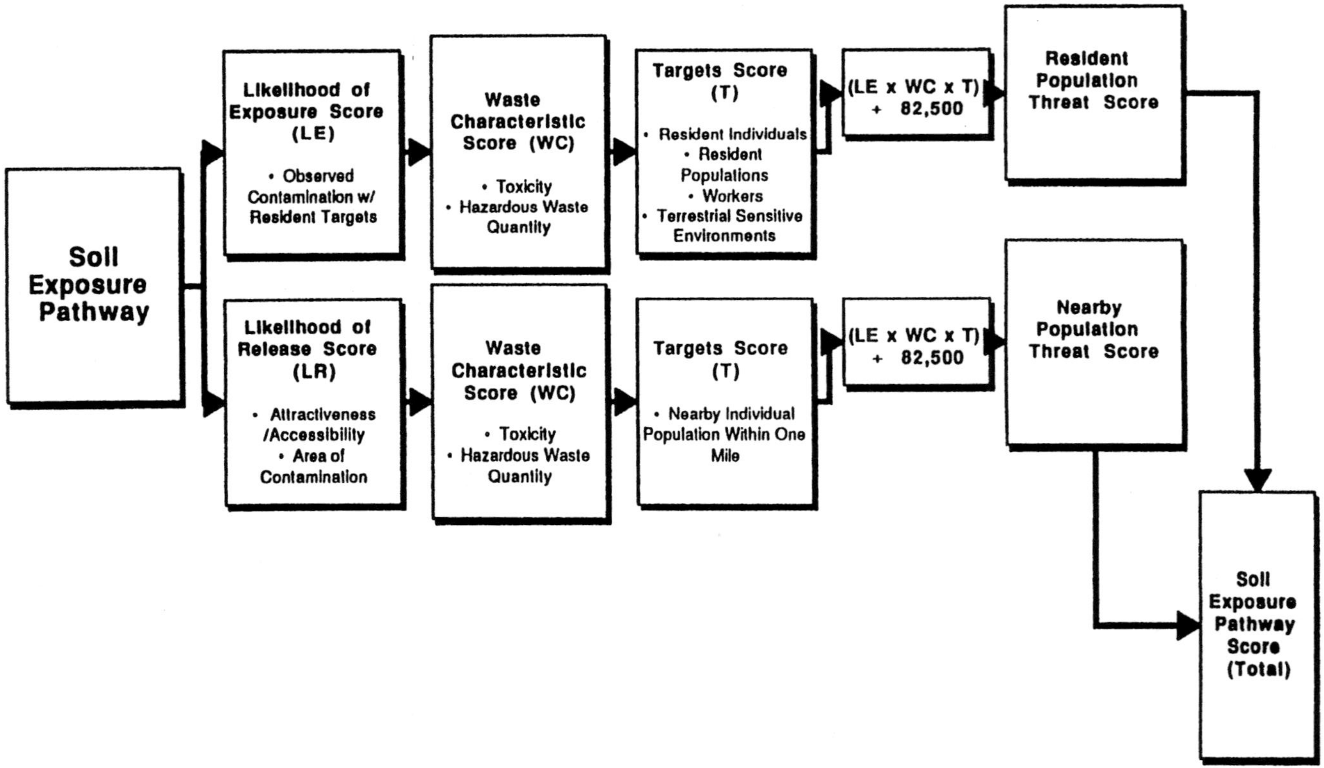

FIGURE 4-3B (CONTINUED)

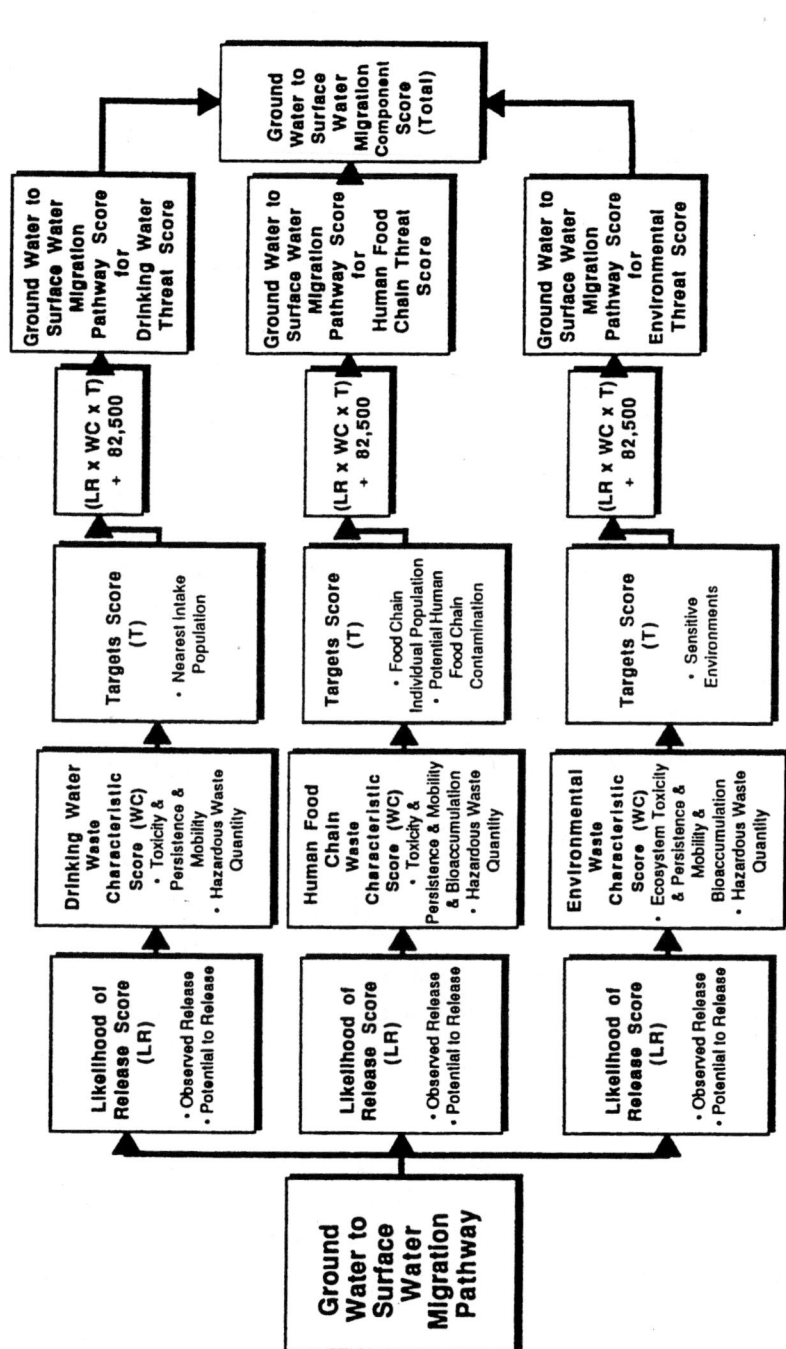

FIGURE 4-3B (CONTINUED)

nearby and on-site impacts on resources and sensitive terrestrial environments. The air migration pathway considers health impacts on nearby populations and environmental impacts on resources and sensitive systems.

Scientific Evaluation of Model Components

The HRS model includes a number of simple analytical and tabular functions for determining the values of individual factor scores. These functions have been developed from a combination of mechanistic factors, empirical relationships and subjective judgment. Given the empirical and subjective character of the model and the ambiguous nature of the overall computed score—intended to provide some index of risk, but not intended to be equivalent to a risk number per se—it is not feasible to perform a rigorous critique of the individual functions and factors that the HRS comprises. Still, it is desired that the relationships reflect good scientific judgment and that a consistent treatment be provided for different pathways and impacts. An evaluation of the basis and consistency of various components of the HRS is provided from this perspective, first with the HRS components common to all pathways and then by a detailed review of the individual pathways. This review considers the HRS as configured by the December 1990 revisions. Previous reviews have been made for the original HRS (e.g., Wu and Hilger, 1984); some of their comments have been addressed in the recent revisions, others remain pertinent.

Likelihood of Release Component

The likelihood of release component accounts for observed releases and the potential for a contaminant to be released. When releases are observed for a particular pathway, the maximum score for that component is assigned directly, eliminating the need for further fate and transport considerations for that pathway. If no releases are observed, the potential for release to that pathway must be estimated.

The potential for contamination of the groundwater is based on the presence or absence of containment, the net precipitation, the depth to the aquifer, and the time it takes for the contaminant to reach the aquifer. The potential for contamination of surface water is based on the possibility of overland flow, which in turn depends on the presence or absence of containment, runoff characteristics of the site, and the distance to the surface water. If the site is in an area that is subject to flooding, then the likelihood of release is also dependent on the presence or absence of containment, and on the flood frequency. In addition, because contaminated groundwater may discharge to the surface water, consideration is given to the influence of containment, net precipitation, depth to the aquifer, and travel time for contaminated groundwater to discharge to the surface water. For the soil exposure pathway, only observed contamination is considered. For the air migration pathway, the potential release is considered for both gaseous and particulate emissions. For gaseous releases, the presence and effectiveness of gas containment measures, the type and source of the gas, and the gas migration potential are evaluated. For particulate releases, the presence and effectiveness of particulate containment measures, the particulate type and source, and the particulate migration potential are evaluated. Each of these considerations is based on site-specific information and does not consider the characteristics of the contaminants.

When a release has not been observed in a pathway and the potential for release is calculated instead, the maximum value for the likelihood of release component is taken as 90% of the value that would have been assigned had a release been observed. Since the HRS is used to rank sites before a full set of environmental data has been collected, it is appropriate to allow a comparable, though slightly smaller, value for the likelihood of release component when the potential for release is high.

Waste-Characteristics Component

The quantity of hazardous substances and their characteristics of toxicity, persistence, and mobility are used to calculate waste-characteristic component scores for each of the four migration pathways. The toxicity factor for the hazardous substance and the hazardous-waste quantity are common to all the pathways, except for the surface water pathway, which treats ecosystem toxicity separately. The persistence and mobility factors are pathway-specific and are considered in the more detailed reviews that follow for the individual pathways.

The toxicity of each substance is derived either from the reference dose data (RfD) for chronic toxins or from weight-of-evidence slope factors for carcinogens. As a fallback position, if no data on chronic or carcinogenic toxicity are available, toxicity factors are determined from a table of LD_{50} (acute toxicity) data for various exposure pathways. A major flaw in this approach is that the primary method of determining the toxicity factor value is independent of the exposure pathway. Human toxicity is dependent on the exposure pathway, and a toxicity database developed for removal actions and other purposes should allow consideration of the exposure pathway. Table 2-4 of the *Federal Register* Final Rule statement for the HRS model (reproduced here as Table 4-1a,b,c) provides ambiguous guidance for toxicity factor evaluation by assigning a value of 0 if RfD and slope data are not available, and also directing the use of the table of acute toxicity in such a case (*Federal Register*, 1990).

The toxicity-persistence-mobility component of the waste-characteristic score for each pathway is calculated for each hazardous

TABLE 4-1A Toxicity Factor Evaluation: Chronic Toxicity (Human)

Reference dose (RfD) (mg/kg-d)	Assigned value
RfD < 0.0005	10,000
0.0005 ≤ RfD < 0.005	1,000
0.005 ≤ RfD < 0.05	100
0.05 ≤ RfD < 0.5	10
0.5 ≤ RfD	1
RfD not available	0

Source: *Federal Register*, 1990.

substance found at the site. The hazardous substance with the highest score is used to calculate the final score for the waste-characteristic component. This procedure, which allows different hazardous substances to contribute to the component scores for different migration pathways, has two flaws. First, the HRS procedure does not give greater scores to sites that have a large number of hazardous substances. Although methods for determining the overall impact of multiple chemicals and chemical mixtures are only in their infancy (e.g., Arcos et al., 1989), it is likely that the cumulative effect of many chemicals at a site will be more harmful than the effect from a single substance. Second, the method does not allows the weighting of hazardous substances based on the amount present. The latter flaw could result in greater scores for sites having a small quantity of a hazardous substance that is slightly more toxic than another hazardous substance found at other sites in much larger quantities. This could be corrected by selecting the hazardous substance used to represent each pathway to be the one yielding the greatest waste characteristic score, rather than the substance with the highest toxicity or combined-factor value. Such a selection is desirable because that score is a result of the product of the toxicity or combined-factor value and the hazardous-waste quantity factor value.

The hazardous-waste quantity factor is determined by estimating the mass of the hazardous substance at the site. For hazardous wastes that are listed for reasons other than toxicity, the entire

TABLE 4-1B Toxicity Factor Evaluation: Carcinogenicity (Human)

Weight-of-evidence[a]/slope factor (mg/kg-day)$^{-1}$			Assigned value
A	B	C	
0.5 ≤ SF[b]	5 ≤ SF	50 ≤ SF	10,000
0.05 ≤ SF < 0.5	0.5 ≤ SF < 5	5 ≤ SF < 50	1,000
SF < 0.05	0.05 ≤ SF < 0.5	0.5 ≤ SF < 5	100
	SF < 0.05	SF < 0.5	10
Slope factor not available	Slope factor not available	Slope factor not available	0

[a]A, B, and C refer to weight-of-evidence categories. Assign substances with a weight-of-evidence category of D (inadequate evidence of carcinogenicity) or E (evidence of lack of carcinogenicity) a value of 0 for carcinogenicity.

[b]Slope factor.

Source: *Federal Register*, 1990.

mass of the waste is estimated. Hazardous-waste streams are arbitrarily assumed to contain 0.02% of a hazardous constituent; waste in landfills is assumed to contain 2×10^{-5}% hazardous constituents, while soil in a land treatment facility is assumed to contain 0.0086% hazardous material. No rationale is evident for these and other values selected for determining the hazardous-waste quantity.

The product of the toxicity factor and the hazardous-waste quantity factor, which is a "hazard-scaled" mass of the substance or waste present, is assigned a second scaling factor according to the $\log_{10}$ of the product. Although this may be appropriate, no rationale for the logarithmic relationship is provided, and this approach might result in an underweighting of sites for which the waste characteristic product is high.

TABLE 4-1C Toxicity Factor Evaluation: Acute Toxicity (Human)[a]

Oral LD_{10} (mg/kg)	Dermal LD_{50} (mg/kg)	Dust or mist LC_{50} (mg/l)	Gas or vapor LC_{50} (ppm)	Assigned value
$LD_{50} < 5$	$LD_{50} < 2$	$LC_{50} < 0.2$	$LC_{50} < 20$	1,000
$5 \leq LD_{50} < 50$	$2 \leq LD_{50} < 20$	$0.2 \leq LC_{50} < 2$	$20 \leq LC_{50} < 200$	100
$50 \leq LD_{50} < 500$	$20 \leq LD_{50} < 200$	$2 \leq LC_{50} < 20$	$200 \leq LC_{50} < 2,000$	10
$500 \leq LD_{50}$	$200 \leq LD_{50}$	$20 \leq LC_{50}$	$2,000 \leq LC_{50}$	1
LD_{50} not available	LD_{50} not available	LC_{50} not available	LC_{50} not available	0

[a] LD_{50} refers to a toxicant dose that is lethal for 50 percent of the test subjects. LC_{50} refers to a toxicant concentration that is lethal for 50% of the test subjects.

Source: *Federal Register*, 1990.

Critique of the HRS Toxicology and Exposure Components

To predict the relative degree of human health hazard posed by chemicals of different toxicity associated with potential NPL sites, the HRS strategy employs toxicity factors computed from regulatory limit concentrations and screening concentrations. Chemical properties such as mobility and persistence are considered with toxicity to describe the waste characteristics. To some extent, these attributes of the HRS model mimic the "exposure assessment" and "dose-response assessment" portions of a quantitative health-risk assessment. However, as demonstrated in this section, the model appears to overemphasize toxicity considerations by assigning only relatively low levels of expected environmental human exposure. An internal inconsistency in the ranking of carcinogens versus noncarcinogens also appears. In most cases, and particularly with regard to potential acute exposures, carcinogens seem to be overweighted in severity of effect compared to noncarcinogens.

The greater the toxicity of chemicals present at a site, the greater the potential for harm. Toxicity factors are thus appropriate for inclusion in the HRS. However, toxicity factors for different health endpoints are not directly comparable. Consider the numerical expression for the toxicity of carcinogens versus that of noncarcinogens. Carcinogens are considered to act by nonthreshold mechanisms, while noncarcinogens are assumed to have thresholds for toxicity. Because both types of contaminants are found at waste sites, it is necessary to compare and weight them for the overall toxicity score.

Different procedures are used for assigning relative weights to carcinogens and noncarcinogens in the HRS. As shown in Table 4-1 (HRS Table 2-4), the reference dose (RfD) is used to score chronic toxicity for noncarcinogens and the slope factor is used to score carcinogens for the waste characteristics factor category. For noncarcinogens with the same score as "B" carcinogens, the RfD is equivalent to a lifetime cancer risk of 7.8×10^{-3}. In the targets factor category, screening concentrations are used as triggers to place observed contaminant concentrations into the Level II or more serious Level I category.

Screening concentrations are concentrations that result in the RfD for noncarcinogens and a 1×10^{-6} risk for carcinogens. The equivalent cancer risk for a noncarcinogen at the RfD is thus nearly four orders of magnitude lower in the targets factor category than in the Waste Characteristics factor category (1×10^{-6} vs. 7.8×10^{-3}). The HRS documentation provides no apparent reason for this inconsistency.

The RfD for a noncarcinogen is the maximum acceptable dose to which a person could be exposed over a lifetime with no ill effects. Similarly, one in one million excess risk of cancer is considered to be a de minimus risk. The use of these gauges for ranking toxic effects of chemicals is most appropriate for doses that are close to the trigger levels. Doses much higher than the trigger levels would produce a different ranking order for hazard because of each chemical's unequal advance toward acute or lethal effects from a low or acceptable dose. In particular, for carcinogens, the dose-response curve is assumed to be linear with a constant slope. One thousand times a risk of 1×10^{-6} is quite far below the background incidence rate of cancer of 25%. One thousand times an RfD on the other hand would be a lethal dose for many noncarcinogens. Thus, evaluations of the toxicity of noncarcinogens should consider the concentration relative to the reference and the lethal dose.

To illustrate the potential distortion produced by the HRS at high dose levels, consider a comparison of cyanide (CN) and cadmium (Cd). Assume 1mg/kg-day doses are generated for CN at one site and for Cd at another; CN is then at 50 times its RfD and Cd is at 2000 times its RfD. This results in assigned toxicity-factor values of 100 and 1000, respectively, showing Cd to be worse than CN. At this dose, CN is at 0.2 times its LD_{50} and Cd at 0.004 times. From this perspective, CN is a much more serious threat than Cd, but its ranking indicates it is 10 times less serious. Although most sites would not produce doses of this magnitude and the ranking system is properly aimed at the possibility of long-term chronic exposure, sites yielding doses significantly above the RfD could be incorrectly ranked.

The toxic properties of chemicals also affect the HRS ranking through the use of regulatory limits and screening levels described in Section 2.5 of the *Federal Register* (1990) Final Rule. The concentration of a chemical detected in the groundwater or surface water at a site is compared to existing regulatory limits such as

Maximum Contaminant Levels (MCLs) or Food and Drug Administration Action Levels (*Federal Register*, 1990, Section 2.5.2). If these are not available, the contaminant concentration is compared to screening concentrations that produce a 1×10^{-6} risk for carcinogens or the RfD for noncarcinogens. Although the regulatory limits listed in the HRS in general reflect toxicity, they also incorporate other factors, such as practical quantification limits and cost of compliance. For carcinogens, this means that an MCL might reflect a risk greater than 1×10^{-6}. An example is chloroform, for which the drinking water MCL is 100 µg/L, representing a cancer risk of 1.7×10^{-5} (based on a unit risk of 1.7×10^{-7} L/µg for chloroform in drinking water). The use of regulatory limits in the HRS can thus produce results quite different from a quantitative risk assessment, which considers only toxicity.

The amount of exposure to a contaminant, as well as the degree of its toxicity, determines the risk of an adverse health effect. Procedures for quantitative health risk assessment consider these aspects separately. The HRS combines toxicity with indices of mobility, persistence, and bioaccumulation. Those factors are multiplied in separate steps of the HRS and then combined in the overall waste characteristics category. Many other attributes of the HRS serve as surrogates for exposure, such as waste quantity, containment, and soil characteristics. The lack of separate subscores indicative of chemical toxicity and of exposure likelihood and magnitude makes it difficult to assess how the HRS conceptually compares to a quantitative risk assessment.

Groundwater Pathway

A groundwater migration pathway subscore is calculated for each aquifer at the site, and the highest of theses is used. Net precipitation factor values are derived from maps or on-site data for annual precipitation, less evapotranspiration. Unfortunately, the method does not take into account runoff, which can significantly reduce the flow rate of water to the subsurface, even for soils with a relatively high infiltration rate. When the map of net precipitation factor values (Fig. 3-2 on page 84 of the final HRS rule) is compared to net percolation depths for soils of Hydrological Groups A, B and C (Brown et al., 1977), it is evident that even though the shapes of some of the zones are similar in some parts of the country, they differ significantly in others. For example, areas of greatest percolation to the groundwater are in eastern Tennessee and Kentucky, not in northern Alabama and Georgia and eastern North Carolina as shown. These discrepancies could result in the erroneous scoring of some sites.

Although one might argue that it is sufficient for the present purpose to use net precipitation without accounting for runoff, such detail is included in other parts of the calculations. Specifically, soil hydrological groups and rainfall runoff values are used in Section 4.1.2.1.2.1.2 of the final HRS rule for the surface water component of the scoring. It would be better to use the same level of detail for calculations throughout the scoring.

The model's value representing likelihood of release to the groundwater includes factors on the depth and travel time to the aquifer. Although they are structured in such a way that the travel time is based on the thickness of the restricting layer and not the entire depth to the aquifer, inclusion of both factors introduces some degree of double counting, depending on the relative thickness of the most restrictive zone. Furthermore, the value of the hydraulic conductivity of 10^{-8} cm/sec assigned to clay and low permeable till (compact unfractured till) in Table 3-6 of the HRS final rule is inappropriate. Such materials often have much greater permeabilities (Freeze and Cherry, 1979; Griffin et al., 1985), and the use of the low conductivity values will result in very low factor scores. Furthermore, at some sites with hazardous wastes, channels for leakage have been created by drilling activities such as for wells and mine shafts. Natural faults allow liquids to flow rapidly through zones that would otherwise be classified as having low permeability.

The mobility of each hazardous substance in groundwater is governed by a number of factors including the water solubility of the substance and its soil sorption coefficient. The model's procedure appropriately results in greater values for karst environments, and higher values if the substance is present as a liquid. However, the partition coefficient, K_d, can only adequately represent partitioning between water-soluble substances and the solid matrix. It is not appropriate to use it when a nonaqueous phase liquid (NAPL) is in direct contact with the soil solids. Thus, it is inappropriate to adjust the groundwater mobility factor values downward from 1 for liquid wastes when the K_d value is >10 ml/g. The value should remain at unity whenever an NAPL is present. To do otherwise will underestimate the potential for mobility and result in an underscoring of sites. The mechanisms that facilitate movement (or, conversely, cause retardation) are quite different for inorganic and organic species. Thus combining K_d and water solubility in an apparent attempt to handle inorganic and organic species in the same table is not justified. K_d is widely used for organic species and calculated as follows: $K_d = (K_{oc}) f_{oc}$, where f_{oc} is the fraction of organic carbon and K_{oc} is the partition coefficient between the water and the organic carbon in the soil (Roy and Griffin, 1989). In general, is not necessary to include the fraction of clay since there is negligible adsorption of organics on the clay fraction if water is present, even for NAPLs.

Solubility only partially controls the movement of inorganic species. Partitioning of cations is regulated by their speciation (valence state, type and degree of organic ligand formation, pH, and oxidation state), the charge density of the medium, and the presence of competing ions. In most of the situations where cations are present the potential concentrations at the receptor points would be well below the solubility of the compound in water. The groundwater mobility factor value does change in the appropriate direction as the K_d changes, but the suggested dependence on solubility is not technically correct and could lead to erroneous scores, particularly when comparing sites with primarily organic contaminants to those with primarily inorganic contaminants.

Surface Water Pathway

The surface water migration pathway considers overland flow and release caused by flooding. The runoff factor value of the overland flow component is derived from the U.S. Department of Agriculture (USDA) curve number (SCS, 1972) using the two-year, 24-hour rainfall frequency. This approach is appropriate but the resulting values are not proportional to the amounts of runoff that are predicted for the soil hydrological groups. The HRS factor values for soil groups A, B, C, and D for the 3.5-inch rainfall can be normalized to be 1, 1.33, 1.66, and 2 respectively. However, actual average runoff amounts generated by the USDA method have ratios of 1, 2, 3.6, and 4.2 respectively. Thus, the HRS procedure undervalues the amount of runoff for the less permeable soils in groups C and D, relative to those in Group A, by a factor of over 2. It is possible that this was done deliberately to account for increased dilution resulting from greater runoff, but this was not documented and is not appropriate since much of the transport of contaminants will be associated with erosion, which increases as runoff increases.

The flood frequency factor values are equal for annual and 10-year flood plains. A small adjustment, perhaps setting the value for the 10-year flood plain 10% lower than that for the annual flood plain, would be consistent with similar distributions assigned elsewhere in the procedure. Persistence factor values for substances in surface water are determined as the greater of the values determined either by the half-life or the logarithm of the octanol-water partition coefficient, log K_{ow}. Although mobility is dependent on the K_{ow} in the environment, and K_{ow} may be related to the relative partitioning of chemicals into the fat in fish (as it is correctly used in calculating the bioaccumulation factor), it is not necessarily related to the rate at which hazardous substances are metabolized. Thus, its inclusion for the purpose of estimating persistence may be inappropriate, e.g., for poorly degradable water-soluble heterocyclics. The bioaccumulation factor values may be arguable whether calculated from actual bioconcentration data or from log of K_{ow}, but they are not related to water solubility. Thus, this portion of Table 4-15 of the final HRS rule might give misleading scores.

Soil Pathway

The soil exposure pathway score is based on the exposure of workers, residents and nearby populations. It also includes a ranking for the sensitivity of terrestrial environments. The scores are based on the size of the contaminated area and the value of the hazardous-waste quantity factor for the selected contaminant. The only pathway considered for soil exposure is direct ingestion by residents, site workers, or other individuals who may visit the site. Inhalation of associated gases or particles are addressed in the air pathway. Dermal exposure is not addressed explicitly. Because exposure is assumed to occur when individuals visit the site, there are no considerations of fate and mobility contributing to the soil pathway score.

Air Pathway

The air migration pathway score is based on the exposure of

individuals or the population within one mile to observed or potential gaseous or particulate releases for the selected hazardous substance. Factor values used in the score for potential releases include containment considerations, source-type considerations, and the vapor pressure and Henry's Law constants for the substance (Table 6-4 of the final HRS rule). The assigned values appear to be internally consistent, and also consistent with our understanding of the importance of each, with the exception of the apparently low factor value assigned for evidence of biogas release from landfills. It is known (Wood and Potter, 1987; Smith et al., 1989) that landfill gases are effective in transporting hazardous substances, and thus this value appears low as compared with others in Table 6-4 of the final HRS rule. The value for potential release from surface impoundments also appears low, since these are direct sources of air emissions.

Particulate migration is based on observed releases or the site-specific mobility factor used to calculate the likelihood of release. Mobility is also a component of the waste characteristic factor value and is determined from site-specific information or ranges provided. Thus, particulate mobility is included in the scoring twice for substances with lower vapor pressures. The maps provided to determine the particulate migration potential factor values and the particulate mobility factor values are different from each other and also differ from maps delineating areas of actual (Plaster, 1985) or potential (Donahue et al., 1977) wind erosion for the continental United States.

Summary Evaluation of Pathway Calculations

The fate and mobility factors of hazardous substances are used to determine potential exposure for three of the four pathways. In most instances, the judgmental values used to evaluate containment appear to be appropriate. Double counting for factors influencing mobility is evident in the groundwater and air migration pathways. Since the score is based on just one substance for each pathway, it is critical that the most appropriate substance be selected. The selection should include consideration of the quantity of the hazardous substances in addition to the toxicity and mobility factors that are now the only controlling factors. Another problem with the scoring methodology is that the toxicity factor is not weighted for the means by which the individual is exposed.

There are differences in the details of the mechanisms controlling mobility in the different pathways. Such differences might or might not have a significant impact on the score, but a consistent level of mechanistic detail should have been used throughout. In a few instances, it is apparent that the judgmental values assigned to site-specific conditions may not be completely reflective of the relative hazards. Examples of this are consideration of the presence of NAPL and the mobility attributable to landfill gas, which are likely undervalued. Although the scoring procedure is generally logical in terms of the direction of effects of different input factors, some of the observed flaws in the fate and transport components of the scores could result in the scores of sites, particularly those with no observed releases, being inappropriately ranked.

Ecological Factors in the HRS

An important aspect of the recent HRS revisions was EPA's desire to "improve the evaluation of sensitive environments by addressing a broader range of sensitive ecosystems and to afford a higher weight for sensitive environmental factors" (Caldwell and Ortiz, 1989). Although sensitive environments were included in the original HRS, a site that had only environmental impacts in the absence of human health effects, even if it had major impacts on an endangered species or a national park, could not score high enough for placement on the NPL.

The current HRS includes impacts on sensitive environments in the surface water, soil, and air migration pathways. Sensitive environments in the surface water pathway are rated from 5 points for state-designated areas for protection or maintenance of aquatic life, to 100 points for critical habitats and other formal federally designated areas. Soil impacts are included for sensitive terrestrial environments, with ratings ranging from 25 for state lands designated for wildlife management to 100 for federally designated critical habitats and endangered species areas. A range of sensitive environments is considered for the air pathway, with special emphasis on large wetland areas.

An important issue in understanding and comparing the different priority-setting and ranking models for hazardous-waste sites is the relative weights applied to the ecological versus human health targets. Are the weights clear and identifiable in the development and presentation of the model, and is an adequate (or for that matter, any) rationale presented for their selection? The weights for environmental impact in the HRS remain lower than those assigned to human health, but are intended to be high enough that sites which seriously threaten an important sensitive environment can score sufficiently high for placement on the NPL. However, the precise weighting in the HRS is difficult to determine because of the various multiplicative and additive steps in the algorithm.

The environmental versus human health weights in the HRS were established using a Delphi method within the EPA workgroup (Caldwell and Ortiz, 1989). Although public comment on the weights was solicited in the preamble to the proposed rule, it is not clear how broadly representative the weights were intended to be, and there are no provisions for adjusting the weights to account for differences in local or regional preferences and values that may be relevant at a site. As discussed in Chapter 9, the committee believes that it would be desirable for a national model to have adjustable value weights to reflect local preferences, perhaps on a state-by-state basis.

Socioeconomic Factors in the HRS

Potential economic benefits from remediating hazardous-waste sites include reduction of community damage, appreciation of property, increased productivity of land, creation of jobs, and reduced expenditures of health care. Social benefits include enhancement of existing communities, especially disadvantaged communities. Although there have been few empirical studies of those benefits, despite their potential importance, recent studies have shown some economic impacts. For example, McClelland et al. (1990) reported that housing values in a Los Angeles neighborhood increased by approximately $5,000 per unit after a landfill was closed. Skaburskis (1989) pointed to a 15% decrease in sale prices near a landfill, but showed no impact beyond 0.25 mile from a site that he described as noncontroversial.

Social impacts of siting and remediating hazardous-waste repositories are also likely to be important, but are difficult to document. Edelstein (1988) interviewed residents of communities with landfills and hazardous-waste sites, attended public meetings, and read meeting transcripts. He concluded that stigmatizing land uses gradually changed people's self-image, image of their family, and images of their community, environment, and government. Adults and children became depressed and pessimistic and felt betrayed by their government. It was also reported that many moved and others wished that their homes would burn down, so that they could afford to move. The implication of this research is that remediation can ameliorate such problems.

Economic effects that are generally evaluated include those involving site and off-site land use and property values, employees, and neighboring populations. For example, remediation of contamination at a site might increase the values of nearby residential, farm, industrial, and commercial properties; encourage nearby land-use investment and development; and positively affect the future livelihood of nearby citizens.

Although the original directive to EPA for establishing criteria and priorities for site remediation specifically indicated inclusion of consideration of public welfare, only health and environmental impacts are directly considered in the HRS. Public welfare is considered in an indirect manner, as it is affected by human health and environmental resource impacts. In this respect, the HRS does consider a reasonably broad and representative set of human and environmental resource targets. The human targets include resident, student, and worker populations, including the individuals nearest the site. However, transient populations are not considered, and no consideration is given to the age, sex, or socioeconomic status of the target group or individual. The environmental resource targets (in addition to sensitive environments) include commercial farming, food preparation, recreational areas, and drinking water supplies.

No attempt is made in the HRS to incorporate direct estimates of economic impact (such as property value losses near the site) or social effects (such as disruption of existing communities or distributional effects). The human population and resource components of the HRS do provide some measure of these, and more quantitative, detailed estimates would probably be difficult to make, certainly at the early assessment stage of the HRS. Still, EPA has made the establishment of close ties and effective communication with local communities an important part of their

Superfund strategy, and some further consideration, such as designation of "sensitive communities" (in parallel with "sensitive environments"), could be appropriate.

Protocol for Model Development and Use

As discussed in Chapters 2 and 3, the quality and acceptability of a hazard-ranking model is affected by the procedures and protocols adhered to in the development and application of the model. Included in this are the procedures for model testing and validation, the determination of the sensitivity of the model results to uncertainties in model inputs and formulation, and the provision of effective quality control mechanisms to ensure proper and consistent application of the model to site scoring. Additional issues in model development and use include the level of field testing and peer and public review of the model prior to its release for general use, the degree of user-friendliness, its transparency, and the quality of the documentation that guides the data collection and scoring steps.

In the development of the original and the revised HRS model, EPA, in conjunction with the MITRE Corporation, undertook extensive field testing programs (Chang et al., 1981; Caldwell and Ortiz, 1989; Zaragoza, 1990). These exercises helped to work out a number of early computational errors and perceived inconsistencies in the model. That effort, combined with extensive opportunities for peer review and public comment provided by EPA, has led to a model formulation that many have accepted as reasonable, though not necessarily ideal (Wu and Hilger, 1984; EPA, 1988; Wilson, 1991; Haness and Warwick, 1991).

The quality of the documentation that describes the HRS model is considered generally adequate. It is written in a straightforward manner and decomposes the scoring steps into multiple independent tasks. However, there has been little guidance for sampling and collecting the data that are input to the HRS, a gap which has allowed a wide range in data collection procedures and corresponding levels of effort to collect data. EPA is in the process of developing a Hazard Ranking System Guidance Document that will address this need for individuals or groups who are scoring sites.

In general, the HRS is structured so that the collection of more data leads to a higher score. Lack of data or uncertainty in inputs tends to skew the results towards lower values (Haness and Warwick, 1991). This "reward" structure for greater data-collection effort may make it possible for interested parties to manipulate the HRS score to meet their underlying objectives. States or communities that wish to keep a site off the NPL can limit the sampling and data- collection effort at that site. Conversely, states or communities that are motivated by economic or political factors to have a site placed on the NPL might continue the sampling effort until enough data is uncovered to push the score above the 28.5 threshold. Indeed, affluent communities might be more able than poor communities to invest in the necessary sampling to accomplish this, yielding a potential for socioeconomic inequity in the site selection process. This phenomenon can limit the ability of the HRS to provide a truly representative ranking based on objective environmental criteria.

The presence of the 28.5 threshold can lead to a pattern of behavior among scorers that further limits the utility of the HRS score for subsequent comparison of sites. Many site evaluators will collect the data necessary to push the HRS score above 28.5, then stop. Once the score passes this threshold, the site will be on the NPL, and there is no need for a further scoring effort. This limits the utility of the HRS score as an indicator of relative risk between sites once they are on the NPL; some sites with scores of 29 or 30 might have received higher scores had not the scorers focused their effort solely on the 28.5 threshold.

A number of validation studies have been performed for the HRS model, in an attempt to relate the HRS score to the results of more formal risk analyses or in-depth expert panel studies of a group of sites (Applied Decisions Analysis, Inc., 1987; OTA, 1989; Doty and Travis, 1990). In these studies, little correlation was found between the HRS score and the more rigorous risk estimates. An underlying assumption in these comparisons is that the full risk assessments or panel studies represent "truth". Given the high degree of uncertainty in risk analysis procedures and the wide range of expert opinion that might pertain to the various complexities of a hazardous-waste site, these measures of truth might themselves be highly uncertain and suspect. Furthermore, in the analysis performed by Doty and Travis (1990), the HRS scores were compared to risk estimates based solely on human health impacts. Still, the referenced studies provide the only current comparisons of HRS scores to more detailed site hazard estimates, and based on these more objective and rigorous measures of risk and threat, it is likely that the HRS does yield a significant number of false positives (sites included on the NPL that should not be included) and false negatives (sites left off the NPL that should be included).

Given the potential for errors in NPL decisions based on the HRS score, an ability to consider the accuracy, precision, and sensitivity of the score should be available. Sensitivity analyses of the HRS model have been performed, although the multiplicative nature of the model precludes the determination of an absolute sensitivity since the response to a particular factor depends on the values assigned to the other factors in the model (Haness and Warwick, 1991). In a sensitivity analysis performed by EPA to assist in the development of the PA method (the simplified version of the HRS used for screening at the preliminary assessment stage intended to provide a conservative first estimate of the subsequent HRS score), the model results for 110 test sites were found to be particularly sensitive to the combined contaminant characteristic score, which incorporates toxicity, mobility, persistence, bioaccumulation potential, and ecotoxicity factors (EPA, 1991c). The setting of these contaminant characteristic factors to their maximum value, which can occur at actual Superfund sites, was the only simplification to the full HRS considered by EPA that resulted in PA method scores being significantly different (higher) at many sites from the HRS scores ultimately determined with PA and SI information.

Although selective sensitivity analyses have been performed, no formal mechanism for sensitivity or uncertainty analysis is provided as part of the regular HRS scoring procedure. The process does provide for a review of scores by EPA headquarters and a quality assurance check by contractor personnel. This review plus the QA efforts help to eliminate major errors and ensure a degree of consistency in the scoring process, and are conducted for all sites with initial scores above 25.0. Although the review and QA efforts are helpful, the small range of underscoring allowed for further review (between 25.0 and 28.5) and the exactness of the final cutoff value (28.5, clearly well beyond the precision of environmental assessment models of any type) dictate that a more formal consideration and allowance for uncertainty should be incorporated as part of the HRS process. Such a recommendation, based on detailed review of sites with a wider range of HRS scores, has been put forward by the U.S. Congress Office of Technology Assessment (OTA) and is considered later in this chapter.

Although recognizing that the HRS cannot provide an accurate absolute measure of environmental threat at the level of a detailed risk assessment, proponents of the HRS have generally found the simplicity of the model to be advantageous for consistent application and transparent evaluation of model results, even by nonexperts. However, early reports indicate that this user-friendliness has been compromised to some extent with the promulgation of the revised HRS, which is considered next.

The Effect of the 1990 HRS Revisions

The revised HRS evolved through an extensive procedure of model development, field testing, peer review, and public comment (EPA, 1988; Caldwell and Ortiz, 1989). The model was modified by eliminating the direct contact pathway and the fire and explosion pathway, and by adding a soil exposure pathway. The eliminated pathways were designed to determine the need for immediate removal (emergency) action, and it was thought that they could be better addressed outside the long-term scope of the NPL process. Additions to the model included

- a new exposure pathway for contact with contaminated soils;
- consideration of chronic noncarcinogenic toxicity;
- expansion of the ecological components of the model, allowing for consideration of a wider range of sensitive environments;
- consideration of the potential for air emissions;
- a groundwater-to-surface water migration pathway;
- use of concentration data in determining waste quantity; and
- higher weights for actual exposure and for potential exposure closer to the site.

Some these changes were intended to allow the revised HRS to correspond more closely to a risk-assessment procedure, as described in Johnson and Zaragoza (1991).

The changes to the HRS have resulted in a tool that is superior to the previous version in both the range of issues considered and the types of input data used. However, this improvement has occurred at the cost of a significant increase in model complexity and the amount of effort required to collect the input data and perform the model evaluation. This increase in complexity and in the amount of resources required has taken the model from the realm of an easy-to-use, accessible tool to one that is significantly more difficult to use. The model can no longer be taught as a lab project for engineering students, and it is no longer possible to teach the use of the model in a workshop for groups of lay citizens interested in scoring a site or checking the scoring of a site performed by their EPA or state officials (L. Greer, National Resources Defense Council, Washington, D.C., pers. comm., 1991). Perhaps because of this, a number of states have recently adopted simpler ranking procedures for their non-NPL sites (see Chapter 7). Although dissatisfaction with the added complexity of the revised HRS might diminish as regular EPA scorers gain experience with the model, it is apparent that the model is now less accessible to other users than was previously the case.

A related concern with the revised HRS is that the improvements in the coverage and rigor with which environmental inputs are evaluated may be lost within the empirical, ad hoc algorithm that underlies the HRS calculation. As such, the additional effort at input data collection (which now approaches that of a full risk assessment) might be wasted when the data are processed through the HRS algorithm. From this perspective, structured-value models are only useful when they are kept simple and transparent. Once they are complicated to the extent that the required level of effort and understanding approaches that for a more rigorous scientific procedure, then arguably the structured-value model should be abandoned in favor of the more rigorous approach.

An important issue that arose in revising the HRS model was whether to modify the 28.5 cutoff value used for inclusion of sites on the NPL. To address this issue, EPA considered a different cutoff score that would be "functionally equivalent" to the 28.5 score in the original HRS, where functional equivalency could be defined based on the following: statistical correlation between the original and revised scores, an equivalent number of sites above the cutoff threshold, or an equivalent risk level for inclusion of sites on the NPL. The criterion of equivalent risk was considered appropriate by most commentators to EPA, including states and industry (Zaragoza, 1990). To support the evaluation, EPA compared original and revised HRS scores at 110 test sites (Zaragoza, 1990). The revised HRS scores tended to be somewhat lower, with fewer of the 110 sites scoring above the 28.5 threshold with the revised model (see Figure 4-4). However, based on a qualitative assessment of risks at selected sites (and given the administrative and legal difficulties that could result from a change in the threshold value), EPA did not feel there was sufficient cause to lower the threshold value to maintain a risk equivalency. The 28.5 threshold for inclusion on the NPL was thus maintained.

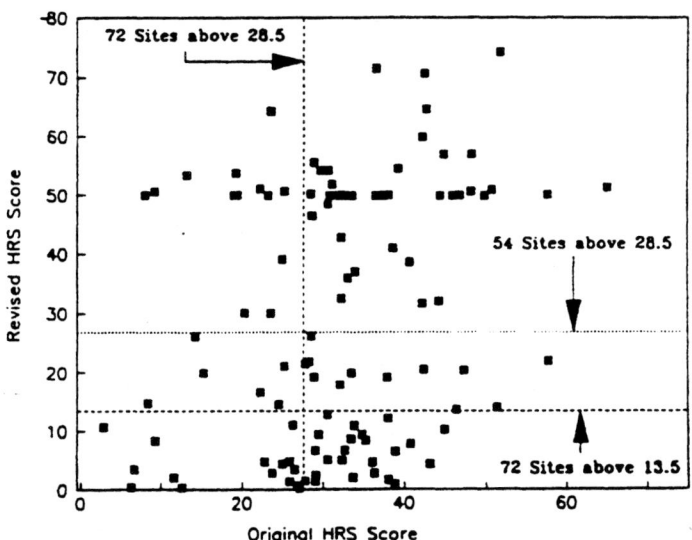

FIGURE 4-4 Scatter plot of site scores for the original and revised HRS. Results were obtained from an EPA study to evaluate the 28.5 cutoff. Source: Wells et al., 1990. Reprinted with permission; copyright 1990, Hazardous Materials Control Research Institute, Rockville, Md.

Priority Setting at Later Stages of Superfund

A backlog of unremediated sites developed as a result of the unexpectedly large number of sites listed on the NPL as well as the significant time and effort required to complete detailed site studies and reach agreement on appropriate plans for remedial action. Such a backlog dictated that EPA develop procedures for priority-setting activities in the later stages of the Superfund process. These procedures include the regional remedial investigation (RI) and feasibility study (FS) priority-setting process and the remedial action (RA) priority-setting process. These processes are not mandated by law and are considerably simpler and less formal than the HRS. However, they do affect the priority assigned to different Superfund sites for remediation and thus need to be considered in this review of alternative priority-setting methods.

The regional RI/FS priority-setting process is not a formal model, but rather a systematic procedure that individual EPA regions must establish to determine priorities for RI/FS projects. The process is applicable to sites and individual operating units and is based on a "worst-first" principle that allocates resources so as to have the "greatest impact on human health and the environment." (EPA, 1990c). The method is applicable only to sites where the costs of the RI/FS could be covered by the Superfund budget; sites where no Superfund dollars are spent, such as federal facility activities or state-initiated enforcements, are exempt from the policy. In addition, other management considerations can be evoked by the regional office to override the evaluation of priority level, including enforcement considerations such as the presence of a willing and financially viable potentially responsible party, or the desire to push forward in-house projects for training, operating unit projects needed for site completion, or projects for sites with multiple, interdependent operating units. State involvement in the priority-setting process is also encouraged.

To implement the worst-first principle, regions are directed to use the HRS package as a starting point and to consider additional information using "standard environmental criteria." These criteria, also used to determine remedial action priorities, include: the risk of contaminants (nature of principal threats), the stability of contaminants, whether human populations are exposed, and threats to significant environments. The result of the priority-setting process is to classify projects which are candidates for RI/FS into three tiers: highest, next-highest, and relatively low priority.

The RA priority-setting process is somewhat more formal, but still quite simple compared with the HRS. Regions determine scores for the standard environmental criteria and for program management considerations, based on questionnaires and a panel review. The scores are combined in a structured-value model, using the weights shown in Table 4-2.

Based on this result, remedial action starts are classified into three categories:

- Priority 1: Immediate or imminent threat.
- Priority 2: Threat from current situation.
- Priority 3: Threat from future situation.

The RA priority-setting process incorporates aggregate, subjective evaluations, but it is simple and quite transparent, so that the reasons for a particular site receiving a given score are clear. This type of model is appropriate for an internal, administrative function, but lacks the formality and replicability of a priority-setting process required by law, such as the NPL selection process wherein the HRS is used.

PROPOSALS FOR IMPROVING SUPERFUND SITE SELECTION AND PRIORITY SETTING

Several suggestions for improving the Superfund site-selection

TABLE 4-2 Weights for Remedial Action Priority Setting

Criteria	Score	Weight	Maximum Score
Risk of contaminants	1-5	x 5	25
Stability	1-5	x 5	25
Population exposed	1-5	x 4	20
Significant environment	1-5	x 3	15
Program management	1-5	x 3	15
			100

Source: D. Evans, EPA, unpublished data presented to the committee, April 10, 1991.

process have been put forth in recent proposals by OTA and other sources. The OTA report, "Coming Clean, Superfund Problems Can Be Solved," (OTA, 1989), notes that many of the features of the site-selection process have been motivated by institutional management constraints rather than environmental or cost-benefit considerations for society as a whole. As mentioned previously, the initial selection of the HRS cutoff value of 28.5 was not based on any inherent environmental-risk threshold or cost-benefit trade-off, but rather the desire at the start of the program to allow an administratively manageable number of sites onto the NPL. The number of sites on the NPL has since grown considerably; however, the HRS cutoff of 28.5 has remained, reflecting the reality that regulatory criteria, once in place, are difficult to change.

Motivated largely by the criticism it received through the early years of Superfund, EPA has modified the Superfund process to encourage quicker progress along the path towards final remediation and closure. The need to demonstrate better administrative progress and control, mandated in part by SARA, has also provided a motivation for EPA to limit the number of sites in the selection pipeline. Fewer sites entering the NPL allow for a better record of progress on those sites that do enter. OTA believes that this motivation has discouraged EPA from undertaking an active-site discovery program, which might lead to hundreds of thousands of sites being placed on CERCLIS, but would avoid future problems that otherwise would occur as these sites are discovered in a delayed and random manner, often by (unpleasant) surprise. Similarly, the motivation to demonstrate progress has encouraged EPA to implement the prescreening evaluations shown in Figure 4-2 (such as the PA method) to further trim the number of sites in the pipeline. Although these decisions are a logical result of the administrative pressures placed on EPA, they do not necessarily encourage environmentally sound decisions. In particular, the strong push to avoid false positives can potentially lead to an increase in the rate of false negatives.

Two of the recommendations put forth by OTA to improve the administration of Superfund directly concern the use of the HRS in site selection. The first is to eliminate the 28.5 score as an exact threshold for inclusion on the NPL. Instead, two HRS scores, one higher and one lower, would be used to classify the candidate sites into three groups. Those above the high score would be selected for immediate inclusion on the NPL. Those below the low score would be deleted as NFRAP cases. Those with scores in the range between the low and high score would then be subject to further review by an expert panel with authority to make the final recommendation for NPL nomination. This procedure would help to eliminate both false positives and false negatives, by allowing more careful evaluation of those sites where selection errors are most likely to occur.

The second OTA recommendation is to combine the PA and SI and RI/FS phase of the Superfund process into a single site-evaluation step, which would then be followed by the HRS scoring. This recommendation, intended to streamline and expedite the overall process, would provide a higher level of site information for use in the HRS scoring. Indeed, as discussed above, many think that the increased data needs of the revised HRS now require information beyond the typical PA and SI effort.

Additional proposals to reform the Superfund process have appeared recently in studies by the MIT Center for Technology, Policy and Industrial Development (MIT, 1992) and by Putnam, Hayes, & Bartlett, Inc. (Butler and Jones, 1992) for the Coalition on Superfund. The MIT study recommends the development of special assessment and remediation procedures for sites with common characteristics, such as landfills, and earlier categorization of sites into immediate action versus no-action but monitor pathways. The Putnam, Hayes, & Bartlett study goes further, and specifically recommends setting priorities for individual actions across sites, rather than setting priorities for the sites themselves, and the identification of early actions that might be taken at sites to significantly reduce risk, even before the NPL decision is made (Butler and Jones, 1992). Both studies recommend that the HRS should be rescored as remediation actions are implemented and that, as warranted, the rescored sites should be deleted from the NPL. The Putnam, Hayes, & Bartlett study specifically recommends that predicted reductions in HRS scores associated with alternative remedial actions be used to assign priority to these actions (Butler and Jones, 1992). This use of HRS score differences to reflect risk reduction benefits runs counter to the committee's scientific assessment that the HRS scores can only be used to reflect ordinal differences in sites, and not cardinal or continuous differences in absolute risks.

EPA has recently responded to the recommendations discussed above as well as others by considering new approaches, such as the Superfund Accelerated Cleanup Model (*Inside E.P.A.*, 1992). That plan would combine the site-screening and risk-assessment studies for the preremedial, removal, and remedial phases into a single study. This change would allow elimination of the distinction between EPA's early removal and long-term remedial programs and encourage expedited progress through the Superfund time frame. Whatever the outcome of this particular proposal, it is likely that further efforts will be made to compress the Superfund timeline. Those efforts, consistent with recommendations of the OTA and others recommendations, suggest that more data could be available when the HRS scoring is performed. Again, the issue arises of whether a structured-value model such as HRS should be used, in contrast to a formal risk assessment, when these additional data are available. However this issue is resolved, current statutory requirements make it probable that the HRS will remain a key part of the EPA priority-setting process for NPL selection.

SUMMARY EVALUATION OF EPA PRIORITY SETTING FOR HAZARDOUS-WASTE SITES

To provide a summary evaluation of the EPA priority-setting process, the evaluation criteria identified in Chapter 2 are examined with the primary focus upon the current (revised) version of the HRS model.

General Issues in HRS Model Development and Application

Clearly Defined Purpose: The HRS model has a well-defined purpose within the Superfund process—site selection for the NPL—and a specified user population made up of those responsible for site scoring. The priority-setting processes for the later stages of Superfund are similarly well defined.

Credibilities and Acceptability: Although certain technical limitations to the HRS model have been identified by the committee and others, including particular aspects of the likelihood of release or exposure category for certain pathways, and questions on the handling of the toxicity component of the waste characteristic category, the committee finds that, in general, the HRS model is (within the context of a structured-value model) generally consistent with accepted scientific understanding and knowledge of the environment. Extensive scientific peer review, public participation, and public comment have been included as part of the model development process.

Appropriate Logic and Implementation of Mathematics: The HRS model includes a combination of additive and multiplicative calculations to obtain pathway and total site scores from the individual factor scores. The calculation of pathway scores as the product of the contaminant release (or exposure), chemical characteristics, and target receptor category scores is patterned according to the multiplicative model for risk. However, the often ad hoc procedures for determining the factor and category scores, and the chemical and site factors that combine source, transport, and exposure-toxicity into single measures, make it difficult to interpret the resulting HRS score in any absolute sense. This problem, endemic to structured-value models, precludes the use of the HRS for evaluating risk-reduction benefits obtained from a proposed or completed remedial activity.

Model Documentation: The documentation for the HRS model is generally adequate, though little guidance has been provided to ensure consistent sampling and collection of input data. This need is being addressed in the HRS guidance document being developed by EPA.

Model Validation: The HRS model has been compared in a number of studies with more detailed site assessments based on risk analysis or expert panels. The degree of correlation with these other estimates has generally been low to modest.

Model Sensitivity and Uncertainty Analysis: A number of studies have been conducted by EPA and others to evaluate the sensitivity of the model to various factors and factor categories. In practice, the scoring outcome is quite sensitive to the overall effort exerted in data collection at the site, with a potential for manipulation of this effort by interested parties.

Specific HRS Technical Features

Applicability to All Waste Sites: The HRS model is broadly applicable to the range of hazardous-waste sites encountered.

Allowance for Dynamic Tracking: Proposals have been made to use the HRS model at various stages of the Superfund process to set priorities and track alternative remedial actions. While this could provide a useful administrative tool, such use is not consistent with the ordinal (non-absolute) nature of the HRS normalized score.

Discrimination between Immediate- and Long-Term Risk: The model is intended for long-term risk (greater than 20 years), because immediate threats have been addressed by EPA prior to the NPL-listing decision. The recent EPA plan to remove the sharp division between immediate response and long-term remediation, if implemented, would dictate the need to reintroduce immediate threats to the HRS model.

Inclusion of Cost Estimates of Remediation: The HRS model does not consider costs or timing issues associated with remediation. These issues are considered in the priority-setting procedures for later stages of Superfund.

Transparency: The original HRS model was relatively transparent, but the recent revisions have made the model significantly more difficult to understand. The effective weights for human health versus ecological impacts are difficult to assess.

User-Friendliness: The model is presented in a straightforward manner that should be relatively easy to follow for regular site scorers. However, the revised HRS is too complex for routine use by lay citizens. The committee is unaware of any interactive computer implementation of the HRS.

Appropriate Security: The hard-copy format for HRS Scoring and the quality-assurance checks provided by EPA and a contractor limit the potential for security problems.

The HRS model has been described as providing a consistent, expedient format for site-scoring and site-setting priorities consistent with legislative mandate (Wu and Hilger, 1984). However, some of the particular features of the model are subject to challenge, and the overall appropriateness of a structured-value approach is questioned, given the availability of risk- assessment procedures. This question might continue to be raised as more detailed site study and evaluation procedures are performed earlier in the Superfund process. To the extent that EPA remains committed to an early decision for inclusion on the NPL, the HRS model may remain the best alternative available. Modifications to allow for more detailed review of sites with intermediate scores could, however, help to reduce the number of false positive and false negative decisions.

Subsequent EPA Priority-Setting Process

The HRS represents a critical, first step for remediation priority setting: deciding which sites to place on the National Priority list. Sites that are placed on this list then are subject to subsequent priority setting to determine which ones to investigate first through the Remedial Action/Feasibility Study (RI/FS) process, following which sites are selected for remedial action (RA). The steps in these priority-setting precesses are well defined and open to public comment and scrutiny, though the selections themselves are generally not open to outside review. The RA process is somewhat formal and involves a structured-value model with weighted consideration of the risk of contaminants, stability, population exposed, significant environments, and program management. The process is well defined, relatively simple, and transparent. It is consistent with an administrative program which is not mandated by law, but still needed to ensure effective management of the Superfund program.

5

DOD'S PRIORITY SETTING

INTRODUCTION

The objective of the Department of Defense Priority Model (DPM) is to aid in the ranking of sites according to their relative threat to human health and the environment. According to DOD (1991b),

The DPM relies on site data gathered during the preliminary assessment/site inspection (PA/SI) and remedial investigation/feasibility study (RI/FS) phases described in the National Contingency Plan (40 CFR 300). The product of the DPM is a normalized score from 0 to 100 that is an indication of the relative risk posed by that site to human health and the environment. It is DOD's policy to accomplish site cleanups on a worst case basis, and the DPM has been used to establish this priority order.

The DPM is a mathematical algorithm or model used to compute a numerical score from 0 to 100 that represents the relative potential threat to human health and the environment posed by a contaminated site. Like EPA's Hazard Ranking System, the DPM is a structured-value model. Using quantitative data and qualitative estimates, the DPM calculates separate subscores for effects on human and ecological receptors via surface water and groundwater pathways and air and soil pathways for volatiles and dust. The subscores are then combined into an overall site score.

Part of the committee's task was to prepare an interim report evaluating the methods, assumptions, and constraints of the DPM. The results of the committee's analysis presented in that report (NRC, 1992) are reflected in this chapter. After completing its analysis, the committee learned that DOD had decided not to use the DPM to aid in ranking sites.

BACKGROUND AND HISTORY

The DPM is an outgrowth of the Hazard Assessment Rating Methodology (HARM) developed by the Air Force in the early 1980s. The HARM included surface water and groundwater pathways and considered contaminants present at a site and the potential for exposure of receptors to these contaminants. The HARM relied on the aggregation of subscores for the pathway-receptor combination by simple averaging. With the subsequent development of the EPA Hazard Ranking System (HRS) (discussed in Chapter 4) to establish eligibility for the National Priorities List (NPL), the HARM was revised to improve the scoring for the surface water and groundwater pathways (e.g., by including floodwater transport, depth to groundwater, and infiltration potential); to improve the use of toxicity information to specifically address the relative potency of each significant contaminant; and to obtain better separation of scores by using a root-mean-square algorithm. The resulting modified model, named HARM II, was developed and first tested in 1986 (Barnthouse et al., 1986). In 1987, HARM II was adopted for DOD-wide use and was renamed the

Defense Priority Model. In response to comments regarding the DPM, it was further modified (*Federal Register*, 1989). The first computerized version of the DPM, known as the automated DPM, was developed in 1988 for use on personal computers running under the DOS operating system. This version was released as DPM Version 2.0 in June 1989. A revised version of the DPM was released in June 1991 and is known as the FY 92 version. A rationale document for the DPM was provided to the committee in October 1991 (DOD, 1991c). Throughout this chapter, references are made to the DPM User's Manual for the FY 92 version (DOD, 1991b).

THE DPM STRUCTURE

Overview

The DPM uses a combination of quantitative data and qualitative approximations intended to rank sites according to their potential threats to human health and the environment. Unlike EPA's HRS, which is used for initial screening, the DPM is used after a remedial investigation and feasibility study (RI/FS) has been conducted and the significance of the contamination at a site has been characterized in detail (see Figure 5-1). Contaminant mass, the number of chemicals and their mobility and toxicity, exposure-pathway assumptions, proximity of receptors, and allowable exposure criteria appear to be the important variables that determine the overall ranking of a site. These factors are combined to mimic risk assessment and yield an overall score. The overall scoring method of the DPM is based on a set of product algorithms that account for the exposure pathway, contaminant hazard, and receptors (human, animal, or plant), similar to the HRS (see Chapter 4):

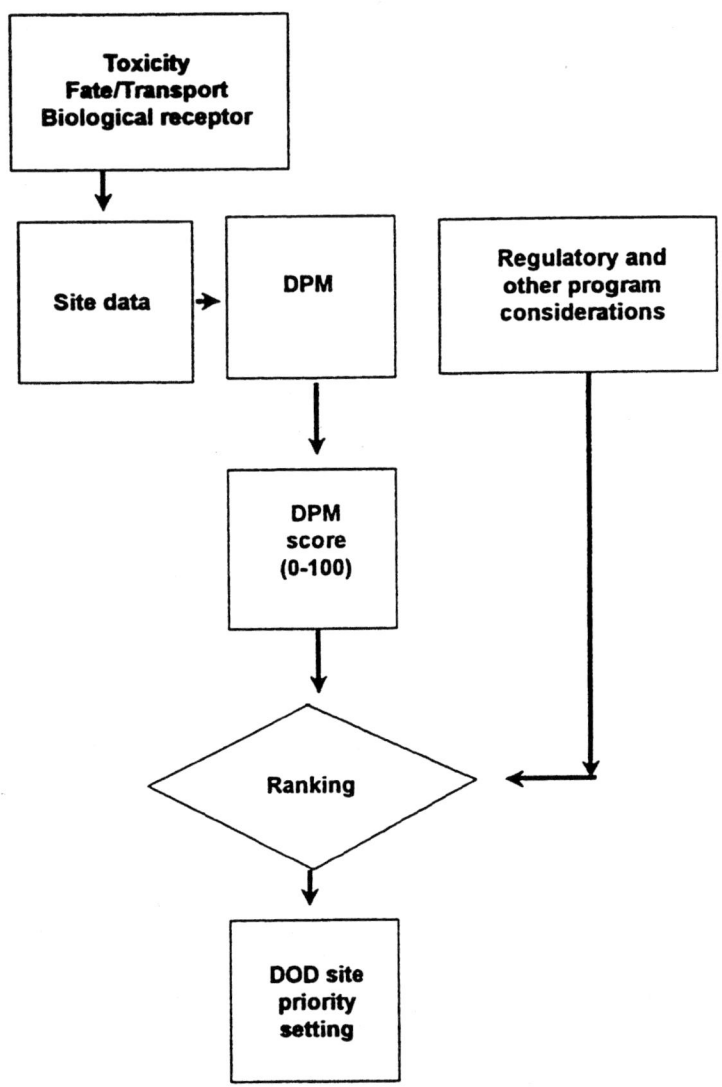

FIGURE 5-1 DOD's approach to priority setting. Source: Adapted from DOD, 1991b.

$$\text{SCORE} = \text{(Pathway Score)} \times \text{(Hazard Score)} \times \text{(Receptor Score)} \quad (5.1)$$

A schematic illustration of the DPM structure, which describes the above components of the DPM, is given in Figures 5-2a and 5-2b and Table 5-1. The final normalized DPM score represents the relative potential threat that a contaminated site poses to human health and the environment. The scores do not constitute the full process of setting priorities for remediation, but they are intended to be one important factor in priority setting (Figure 5-1). The scores can be used by DOD decision-makers with regulatory considerations, program efficiencies, additional risk information, and other factors to determine the relative priority of sites for remedial action (DOD, 1991b).

Data Requirements

The DPM input data requirements are summarized in Table 5-2; the preference is for site-specific data. The information available for DPM application includes data collected as part of the PA and SI and RI/FS activities. To incorporate a measure of uncertainty in the scoring during the quality assurance (QA) assessment, the DPM uses the concept of confidence factors. These confidence factors, which range from a value of 0 (uncertain) to 1 (certain), are multiplied by the maximum possible score for each factor (DOD, 1991b).

Model Documentation And Software

The main documentation describing the DPM is the DPM

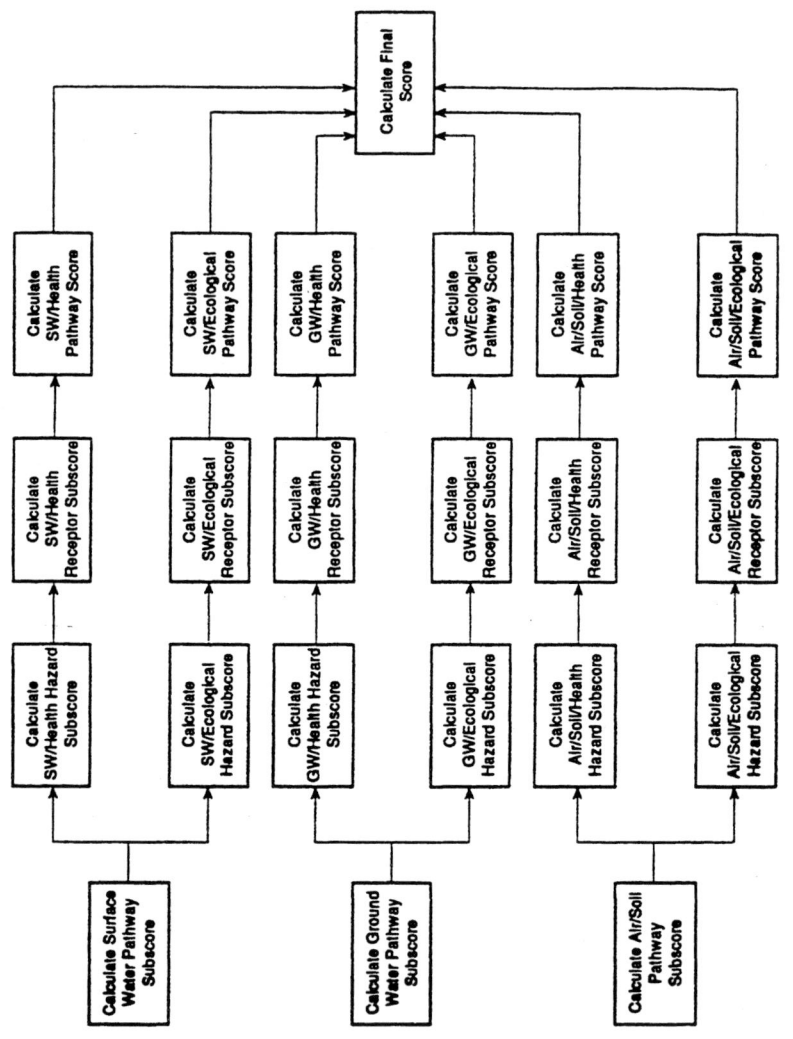

FIGURE 5-2A DPM structure. Source: DOD, 1991b.

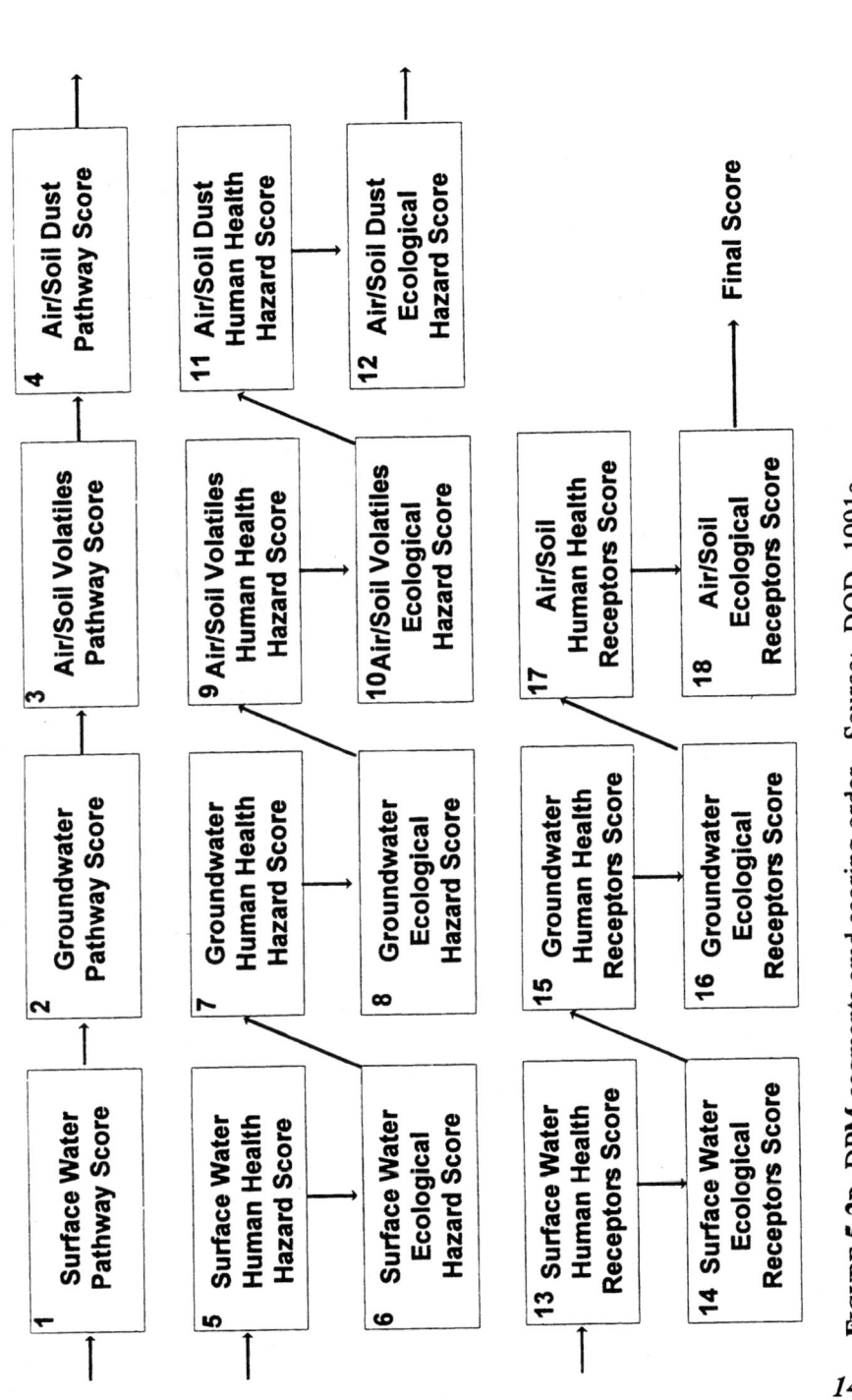

FIGURE 5-2B DPM segments and scoring order. Source: DOD, 1991c.

TABLE 5-1 Combining scores in the DPM

Surface water human health score	= Surface water pathway score	x Surface water human health hazard score	x Surface water human receptor score	/10,000
Surface water ecological score	= Surface water pathway score	x Surface water ecological hazard score	x Surface water ecological receptor score	/10,000
Groundwater human health score	= Groundwater pathway score	x Groundwater human health hazard score	x Groundwater human receptor score	/10,000
Groundwater ecological score	= Groundwater pathway score	x Groundwater ecological hazard score	x Groundwater ecological receptor score	/10,000
Air/soil volatiles[1] human health score	= Air/soil volatiles pathway score	x Air/soil volatiles human health hazard score	x Air/soil volatiles human receptor score	/10,000
Air/soil volatiles ecological score	= Air/soil volatiles pathway score	x Air/soil volatiles ecological hazard score	x Air/soil volatiles ecological receptor score	/10,000
Air/soil dust[1] human health score	= Air/soil dust pathway score	x Air/soil dust human health hazard score	x Air/soil dust human receptor score	/10,000
Air/soil dust[2] ecological score	= Air/soil dust pathway score	x Air/soil dust ecological hazard score	x Air/soil dust ecological receptor score	/10,000

[1]The higher of these two scores is used in the final computation.
[2]The higher of these two scores is used in the final computation.

Source: DOD, 1991c.

TABLE 5-2. Summary of DPM Data Requirements

Data Category	Specific Data Needs
General Site Information	Site location, name and number
	Distance to nearest installation boundary
	Distance to nearest residential, industrial, or commercial land use
	Type of waste site, size of waste site (area and depth), type of waste and concentration of contaminant in waste, physical form of waste, waste quantity, and waste containment information
	Mean summer temperature, soil porosity, average wind speed, soil bulk density
Pollutant Field Data	Have non-volatile (vapor pressure less than 10^{-3} mm Hg) or volatile pollutants detected or releases been observed in the ambient air?
	(For observed releases to air identify contaminants and measure or calculate maximum concentration levels; for non observed releases to air but observed releases in surface soil, identify contaminants and the maximum observed concentrations
	Have volatile or non-volatile pollutants been observed in the surface soil?
	(For observed releases of non-volatile pollutants determine (via measurements or appropriate models) the maximum concentration levels in air and surface soil for each contaminant. For non observed releases of non-volatile pollutants to air but observed releases in surface soil, identify contaminants and determine (via appropriate models) the maximum concentration levels in air and soil)
	Have pollutants been observed in ground water?
	(For observed releases identify contaminants and determine maximum observed concentrations)

TABLE 5-2 (continued)

Data Category	Specific Data Needs
Pollutant Field Data (continued)	Have pollutants been observed in surface water? (For observed surface water contamination, identify contaminants and determine (maximum observed) concentrations).
Climate	Net precipitation and representative rainfall intensity east and west of the Mississippi River Days per year with precipitation greater than 0.25 mm Flooding potential Annual average wind velocity
Soil Data	Soil porosity Soil permeability Erosion potential Neutralization capacity (based on soil chemistry) Organic content pH Average summer soil temperature
Groundwater	Depth (seasonal high) "Short circuit potential" (i.e., presence of faults, cracks, etc.) Distance (downgradient) from waste to supply wells, surface water, habitat or natural areas. Hydraulic conductivity of aquifer, effective porosity of soil, hydraulic gradient
Surface Water	Distance from waste site to nearest surface water Water use of the nearest surface water bodies

TABLE 5-2 (continued)

Population and Ecological Information	Sensitive ecosystems located within 4 miles downstream or 1.5 miles radius Presence of critical environments (e.g., habitat or endangered species, nature preserve, wilderness area, important natural resource, etc.) Population downstream that obtains drinking water from potentially affected surface water or groundwater Population within 1/2 mile of the site and population within a 4 mile radius Land use

Source: Material from DOD, 1991b.

User's Manual (DOD, 1991b) which provides instructions on how to run the DPM but does not explain the science or rationale for the various model algorithms. DPM documentation does not detail the whole modeling process (objectives, assumptions, intended use, environment, etc.), or its product. Thus, it is unclear why and how DPM's scores for potential threats are combined with the threat's magnitude, immediacy, and probability. The above information is necessary for evaluating whether a particular type of risk is being quantified consistently and checking whether the model's default values are chosen consistently with some explicitly stated policy.

The DPM software is well conceived, and it can be used with relative ease. The DPM has some internal checks to prevent the input of unrealistic data, such as entering negative numbers where only positive numbers apply, but it does not check for overall physical consistency of the data prior to execution.

Pathway Scoring

Overview

The pathway portion of the DPM rates the potential for waste-site contaminants to enter surface water, groundwater, and the air and soil. The pathway scoring methodology is based on various concepts of contaminant transport and fate. Contaminant transport and fate are modeled qualitatively in the DPM, with the exception of the air and soil pathways. Thus, for the surface water and groundwater pathways, the algorithm does not consider the magnitude of the pollutant release rates or concentrations at the receptor site in arriving at the pathway scores.

Environmental Transport And Fate

The phrase "environmental transport and fate of chemicals" denotes the set of physicochemical and biological processes in nature that determine the exposure pathways, rates, and concentrations from contaminant source to receptor. Verified mathematical algorithms that link source and receptor are derived on the basis of multimedia chemical mobility and reaction phenomena in nature. Aspects of transport and fate are embedded in the DPM, specifically in the exposure pathway and contaminant-hazard factors.

A detailed review of the DPM revealed that the algorithms used for the surface water, groundwater, and air and soil pathway factors are not entirely consistent with accepted theory. The pathway algorithm uses a summation formula, whereas theory suggests that a multiplicative formula or summation on a logarithmic scale would be the preferable approach for scoring the pathway potential.

The DPM makes use of pathway algorithms that attempt to qualitatively capture the dependence of the contaminant concentration, at a given distance from the source, on various physicochemical and transport parameters. The DPM method for scoring the volatiles in the soil and air pathways is discussed below, followed by an alternative example illustrating that scoring algorithms can be developed along a defensible, theoretical approach. This example is then compared with the DPM approach.

The intention of the soil and air pathway analysis is to assess the potential of a pathway link via air from source to receptor. The three types of sources to consider are volatiles from soil surfaces, volatiles from surface impoundments (water surfaces), and dust from surface soils. The transport phenomena characteristics describing each are different. Any one taken alone should generate a potential score. All three together should have an equal or higher score than any one pathway. If contaminants have been detected in air and volatile compounds found in surface soil or impoundments, the assigned normalized score is the maximum score of 10. If contaminants have not been detected, the scoring is assigned through pathway characteristics (see Figure 5-3). The factors listed in Table 5-3 are used in the soil and air volatiles pathway algorithm.

In the DPM, the following scoring algorithm is used:

$$\text{Score} = [12d^{(s)} + 2T^{(s)} + 2R_T^{(s)} + 2V^{(s)} + 2\epsilon^{(s)}] \times [(W_c + W_q)/2](100/60), \tag{5.2}$$

in which the individual scores for each contaminant (designated with the superscript s) vary with the pathway factors as given below:

$$d^{(s)} \propto d$$

$$T^{(s)} \propto T$$

$$(R_T)^{(s)} \propto (R_T)^{-1}$$

$$V^{(s)} \propto V$$

$$\epsilon^{(s)} \propto \epsilon$$

The main difficulty with the above approach is that the rationale for the coefficients in Equation 5.2 and the assigned maximum possible scores is unclear. Moreover, the algorithm does not seem to be based on a recognized physical description of contaminant volatilization.

Various alternatives can be based on a physical description of

DOD's Priority Setting

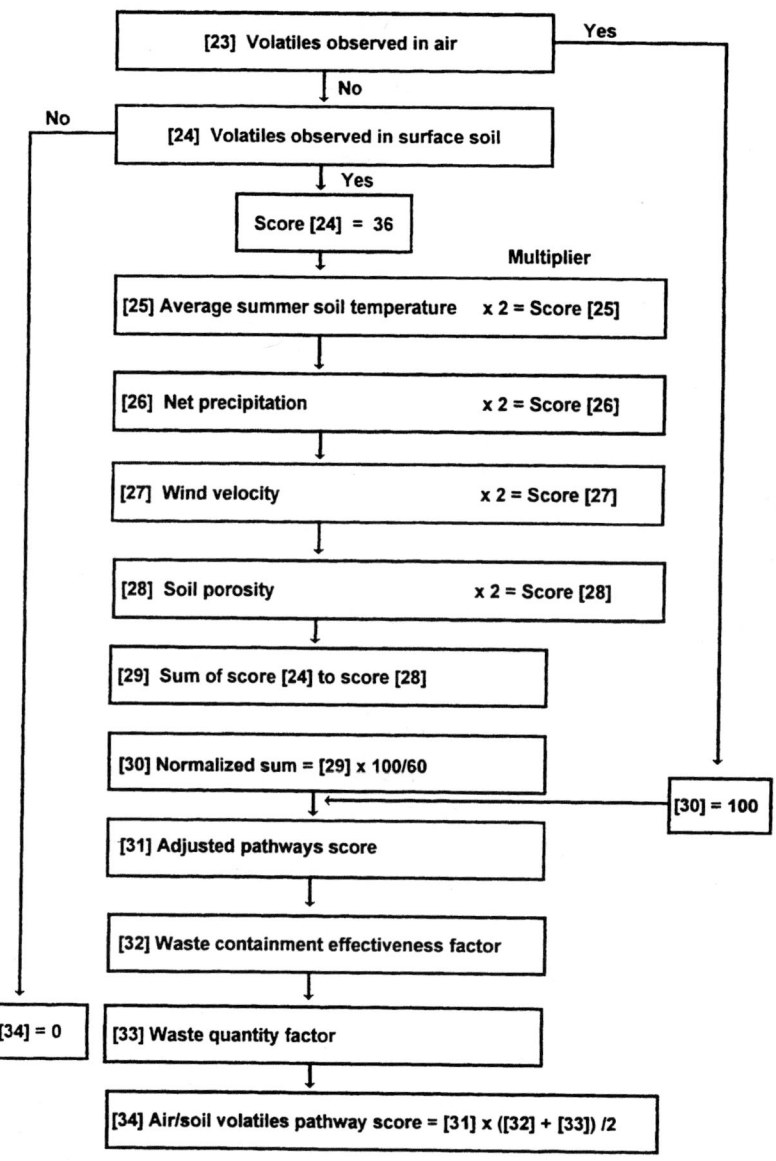

FIGURE 5-3 Air/soil volatiles pathway scoring sequence. Source: DOD, 1991b.

TABLE 5-3 Factors Used in the Soil and Air Volatiles Pathway Algorithm for the Defense Priority Model

Factor	Maximum Possible Score
Average soil temperature, T, °C	6
Net precipitation, R_T, mm H_2O	6
Wind velocity, V, mile/hr	6
Soil porosity, fraction of air space, ϵ, unitless	6
Degree of detected contamination in surface soil, d, unitless	36
Waste containment effectiveness factor, W_c	1
Waste quantity factor, W_q	1

Source: Material from DOD, 1991b.

transport phenomena. An illustration of a scoring algorithm that can be based on a simple treatment of transport phenomena follows. As a starting point, the development of a pathway potential model should focus on the probability of delivering a quantity of contaminated air (i.e., contaminant concentration) from all sources on the site. Concentration, C (g/cm^3), is generated by a flux, F (g/cm^2), from a source of area, A (cm^2), with a wind velocity, V (cm/s), through a boundary-mixing height, H (cm), and site width, W (cm). From a simple mass balance, the chemical concentration in the air phase can be approximated as

$$C = [(F)(A)]/[(V)(H)(W)]. \tag{5.3}$$

Pathway characteristics in this algorithm that do not appear in the DPM User's Manual are the source surface area A, and its width, W. The flux, F, will be different for each source type. In the case of surface-soil volatiles, the flux can be estimated from

$$F = (K)(C_s), \quad (5.4)$$

where K is the transport coefficient and C_s is the concentration of chemical in the surface soil. Substituting Equation 5.4 into Equation 5.3 yields the soil-volatiles algorithm,

$$C = (K)(C_s)(A)/(V)(H)(W). \quad (5.5)$$

The scores in each case are assigned so that they are directly proportional to the pathways factors. A summation algorithm could be devised by working in a logarithmic scale (such as taking the natural log of each side of the equation). Of course, the appropriate mathematical lower limit for the scores should be set to be consistent with the use of a logarithmic scale. Also, waste containment and quantity factors can be added as indicated in Equation 5.2.

Equation 5.5 represents a physically realistic approximation of the dependence of concentration on the pertinent variables. The characteristics of some of the variables in Equation 5.5 are functions of the mode of contaminant volatilization, as described below.

For volatilization from soils, the gas-phase mass-transfer coefficient K is proportional to the wind speed to the 0.8 power (Thibodeaux, 1979). With the wind velocity in the denominator (Equation 5.5), the net effect is $C \sim V^{-0.2}$ – a somewhat weak, but correct effect, in that dispersion at low wind speeds will result in a higher concentration of contaminants at the receptor site (provided that the wind is blowing toward the receptor). If winds are high, dilution occurs, and more of the chemical will migrate upward from the surface and less will migrate laterally to a downstream receptor. The DPM approach predicts a reverse dependence on wind speed.

The concentration of volatiles in air at the soil and atmosphere interface, C_s, can be obtained from a solution of the diffusion equation where the functional dependence of C_s on temperature can be assessed (Ryan and Cohen, 1989). In the limiting case, when the contaminant concentration is high (i.e., saturated phase), a conservative estimate of the void-space concentration can be obtained from knowledge of the saturation vapor pressure (DOD, 1991b, p. I-8). In this case, vapor pressure is an exponential function of temperature (Reid et al., 1977) where, for most volatile chemicals, the natural log of vapor pressure is inversely proportional to the absolute temperature. Thus, it might be more appropriate to construct an exponential scale of $1/T$ with the scores assigned at selected ranges. Alternatively, if a logarithmic scale for scoring Equation 5.5 is selected, then a scale inversely proportional to temperature can be set. The surface area of the source, A, the width of the source, W, and the mixing height, H, should be included as characteristic variables and scored accordingly.

In the case of volatiles from surface impoundments, the flux relationship is also given by Equation 5.4, where C_s represents the interfacial concentration. In this case, K is proportional to wind velocity; thus, C in Equation 5.3 is directly proportional to V. It should be noted that evaporation of contaminants from solutions is not directly proportional to the chemical vapor pressure. This is especially important for aqueous waste containing sparingly water soluble organics; C_s in the case of aqueous waste could be set to the interfacial, liquid-side concentration. This concentration is a function of the bulk chemical concentration in the aqueous phase, mass transfer coefficients for the liquid side and gas side, Henry's Law constant for the chemical, and the bulk air-phase concentration of the chemical. Finally, the characteristics A, H, and W are identical with those for volatiles from surface soils.

Although Equation 5.2 considers net precipitation as a factor, there is no justification for using net precipitation in the soil-volatiles algorithm. Evidence, based on both theory and data, suggests that under some conditions the presence of moisture in soils can lead to an increase in volatilization of some contaminants from the soil. Moisture and porosity are such transient characteristics that obtaining a physically realistic score for their effect on volatilization, given the present state of the art, is highly doubtful. Although the incorporation of precipitation in the groundwater and surface water pathways is appropriate, it should not be incorporated into the soil-volatiles algorithm because it will negate, in part, the scoring for the groundwater pathway.

In summary, the above analysis demonstrates that use of a physical basis for the air-soil volatiles pathway can lead to a scoring algorithm that is very different from the one used in the DPM. This example emphasizes that, whenever possible, the algorithms for pathway scoring should be based on the functional equations for either contaminant flux or for the concentration of the contaminant in the exposure medium.

CONTAMINANT-HAZARD SCORING

Overview

The DPM methodology rates human health hazards and ecological hazards for specific chemicals on the basis of effects benchmarks. In media in which contamination has been detected, the hazard-scoring method is based on chemical intake by human receptors relative to an acceptable daily contaminant intake. For media in which a contaminant has not been detected, but whose presence in the site has been detected in other media, the contaminant-hazard score is based on toxicity factors and bioaccumulation factors. Of the three pathways–surface water, groundwater, and air and soil, the DPM calculates contaminant concentrations with a model only for the air and soil pathway when contaminants have not been observed. For surface water and groundwater, it is assumed that if contamination has been detected in other media, hazard scoring can proceed without knowledge of the contaminant concentration in them.

The contaminant-hazard score is assigned based on the values of hazard quotients defined as (1) the daily intake or acceptable daily intake of the contaminant or (2) the ratio of measured concentration or ecological effects benchmark.

Reference Dose

The reference dose (RfD) of a toxicant is incorporated into the DPM for the ranking of chemicals by degree of chemical hazard. RfD is defined as the amount of a substance to which an individual can be exposed on a daily basis over an extended period (usually a lifetime) without appreciable risk of deleterious noncancer effects (DOD, 1991c). Although that is a common approach, the ranking would be considerably modified if the contribution of other sources of a particular chemical were taken into account in addition to the site-delivered dose. For example, about 70% of the RfD of cadmium is found in a normal human diet. For other toxicants, the dietary percentage contribution to the RfD is considerably less. The harm posed by chemicals from a site is more closely related to the RfD minus contribution from other sources, such as diet, than to the RfD alone.

Bioaccumulation Factors and Health Effects Benchmarks

The use of bioaccumulation factors (BFs) in the determination of health-hazard quotients for observed releases in surface water and groundwater might produce profound distortions in the estimate of true health risk from chemical contamination of these media.

The list of BFs, which range from 10^{-1} to 10^4, is based on ratios of contaminant concentration in an organism to contaminant concentration in water. When empirical data are unavailable, a BF is derived from the chemical properties of the contaminant. The health-hazard quotient is derived from the combination of the BF and the oral health-effects benchmark (HB) and is supposed to indicate the severity of the health risk. The HB represents the toxicity of the contaminant.

BFs derived from measured fish and water concentrations vary widely for a given chemical because the steady-state concentration of a contaminant in a fish is almost never achieved as a result of exposure to the contaminant in water alone, but results from the combined effects of aqueous phase, food-chain, and sediment inputs to the fish. BFs for the same chemical vary widely, depending on different habitats. For example, bottom-dwelling fish, such as eels, accumulate more contaminants than non-bottom-dwelling fish, even though the exposure through water contaminants is the same for both species.

BFs derived from chemical properties, such as the ability to accumulate in fat as estimated by the octanol-water partition coefficient (K_{ow}), must be used cautiously since the DPM procedure does not consider biological responses to contaminants that greatly affect uptake and site of storage. In general, essential nutrients such as manganese, zinc, copper, and selenium are actively taken up from trace amounts in the environment until physiologically appropriate concentrations are reached. Toxic nonessential chemicals, in contrast, might be unregulated and accumulate to an undetermined extent by dissolution in fat or as a substitute for essential elements, such as calcium. PCBs and chlorinated hydrocarbon pesticides accumulate in fat; lead and strontium accumulate in bone, because of their chemical similarity to calcium. Excess exposure to nutrients is usually accompanied by increased metabolism and excretion, so concentrations are maintained within a relatively narrow range. Excretion of toxic substances can be nearly complete, as in the case of some volatile organic chemicals, or nearly absent, as in the case of lead and cadmium.

Toxic substances such as benzo[a]pyrene (BaP) and PCBs can have similar chemically predicated BFs (10^4 for both in the DPM User's Manual, Appendix B; DOD, 1991b), but BaP is metabolized and excreted to a much greater extent than PCBs. Biological mechanisms that result in specific storage sites in an animal create uncertainty about the interpretation of bioaccumulation with respect to human health effects and thus might lead to very poor correspondence between the predicted and actual relationship of bioaccumulation to human health. For example, cadmium accumulates in the liver and kidney but not in muscle tissue. Therefore, if only fish filets are consumed, there is much less risk from cadmium than from contaminants such as mercury, which accumulates in muscle.

Although there is considerable uncertainty in the derivation of some HBs, it does not approach the level of difficulty imposed by ignoring fundamental biological processes, as in the derivation of the BFs. The potential for misranking contaminants at wastes sites due to the use of ill-considered BFs is illustrated by a comparison of zinc (Zn) and cyanide. The calculation of a health-hazard quotient in Column 9 for the surface water hazard with observed releases (DPM User's Manual, Appendix A; DOD, 1991b) involves a division of the total human daily intake by the HB. The measure of toxicity (the HB) for Zn is 10 times that for cyanide. The higher the HB, the less the toxicity. Therefore, for Zn and cyanide, the HBs truly reflect the greater toxicity of cyanide. A problem arises with the consideration of the bioaccumulation potential. Zn has a BF 100,000 times greater than that of cyanide. The origin of the factors is not given in the DPM manual or other supporting documents, but the large difference might be due to Zn being an essential nutrient and reaching much higher concentrations in tissue than in external water because of physiological requirements. Zn is not likely to concentrate in tissues beyond a physiologically appropriate point. Therefore, contaminated environments will not result in the predicted bioaccumulation of Zn if more than trace concentrations are present.

Consider a hypothetical contamination situation in which Zn at 5 ppm (5 mg/L, 5,000 µg/liter) is found in the surface water of Site A, cyanide at 5 ppm is found in the surface water of Site B, and the surface water at each site is a source of drinking water, as is assumed in the DPM. The cyanide contamination would be much more serious because it constitutes a lethal dose for children (Ellenhorn and Barceloux, 1988). The Zn would constitute about the normal daily dietary intake and therefore does not present a problem. However, because of their different BFs, the effect of the total human daily intake for Zn is greatly magnified by the DPM relative to that of cyanide. The result is that the health-hazard quotient is 32 for Zn and 7 for cyanide. Thus, the DPM treats the Zn contamination at Site A as about 5 times as threatening as the cyanide contamination at Site B, whereas the cyanide would be lethal and the Zn would represent no more adverse exposure than that in the normal daily diet.

A more mysterious aspect of the DPM approach is the inclusion of the "bioaccumulation fish factor" in the Groundwater Hazard Worksheet (Observed Releases). Fish are not found in groundwater per se and surface water is considered separately, so there is no reason to consider bioaccumulation in the groundwater pathway. The only human exposure would be through drinking water and other water uses.[1]

---

[1] For the FY 94 version of the DPM, consideration was given to omitting the BF from the calculations and omitting fish ingestion from the ground-

The committee finds that the inclusion of the BF in the calculation of the surface water and groundwater observed release health-hazard quotients could produce an index of hazard less accurate than the use of the HB alone. Except for certain pesticides, PCBs, mercury, a few radionuclides, and sometimes cadmium, bioaccumulation has seldom been a problem in environmental contamination episodes. The formulaic application of measured and derived BFs to hundreds of contaminants probably decreases the overall reliability of the ranking. It would be better to select the few substances that are known to present an increased health risk through bioaccumulation (e.g., those with log BF greater than 3 or log K_{ow} >3.5) and have the presence of these substances trigger a more detailed procedure of the ranking system.

Ecotoxicological Concerns

The current ecological scoring process emphasizes aquatic receptors and is relatively unresponsive to broader environmental threats to vegetation and terrestrial ecosystems. For ecotoxicological concerns, the hazard scores for all but 45 chemicals are greater for human beings than for all other species. Of the 45, only 7 differ by more than 1 unit in favor of ecological receptors. Only 51 chemicals have scores over 1 for ecology, whereas approximately 68 chemicals score 1 or more for health. These figures suggest that there is a systematic bias in toxicity factors that practically ensures that even the most critical ecologically scored sites will not necessarily have a score meriting priority consideration.

Some chemicals are not included but should be in light of known properties and past uses. For example, the absence of 2,4,5-trichlorophenoxyacetic acid and its derivatives gives the im-

---

water hazard assessment. (P. Meehan, DOD, pers. comm., November 27, 1992).

pression that these chemicals are not present in military installations. The wide range of triaryl and mixed aryl-alkyl phosphate esters (used as brake fluids, flame retardants, insulating agent in fabric on wiring, etc.) is covered by a single entry for "tricresyl phosphate." Only a few pesticides are listed, although all military installations had formal discretionary use authority through federal preemption of state and local regulations. Many items that were on the Air Force list of chemicals of concern in an Request for Proposal put out by AFOSR in 1989 do not appear in Appendix B of the DPM User's Manual (DOD, 1991b).

The treatment of terrestrial ecological effects in the DPM is deficient. The only water-quality resource is an outdated compilation in *Water Quality Criteria, 1972* (NRC, 1973). For plants, the PHYTOTOX database (DOD, 1991b, p. H-12) is recommended for its high quality and currency; however, there is no indication that relatively sensitive avian species are considered. Instances in which terrestrial species (especially mammals) are more susceptible or more readily affected than their prey species might be ignored by assessors as for mink with PCBs (Foley et al., 1988). Mitigating factors in the consideration of ecotoxicological effects are the generally higher resistance of mammals than of other vertebrates and certain invertebrates to acute toxic insult, the general absence of better compilations of data, and the historical imperative that 98% of environmental research funded by the federal government has been devoted to aquatic issues. It is hard to hold the DPM developers wholly accountable for not considering the seven basic ecotoxicologic features (Neuhold and Ruggerio, 1976) in the massive screening and scoring exercise that DOD undertakes.

Within ecological risk assessment, a number of concerns can emerge and require attention. For example, loss of economic or highly visible species (such as rainbow trout and largemouth bass) is more likely to attract action than a shift in benthic communities from a diverse mixture of amphipods and gastropods to large numbers of a few species of tubificid worms, although that shift is now widely recognized as constituting important deterioration.

159

However, since health risk scores are weighted preferentially relative to ecological risks, any differences between health and ecological risks are practically irrelevant. Other detailed concerns raised by the committee regarding the treatment of ecological risks in the DPM are described in a separate report (NRC, 1992).

RECEPTOR SCORING

Human Receptor

The DPM contains an informed bias for a large population that is using a water source contaminated by a site close to the installation boundary (DOD, 1991b, pp. 77-79). For example, the high scores for a population near the site could be reduced substantially if, on discovery of the contamination, an alternative water supply were provided. Selectively assigning a lower remediation score to a site in which a particular corrective action (interim remedial measures) has been taken raises a question of its fairness. Obviously, provision of an alternative water supply only accounts for the consideration of human health risk and ignores needs for environmental restoration.

The land use and zoning factor used in the DPM seems redundant with respect to the example above. This factor also neglects the higher probability of contact by agrarian populations. Indeed, the contribution of pollution by a local industrial or commercial complex might be higher than that of the site itself. About the only feature this factor adds is the transient population of workers and customers who might be exposed in the commercial and industrial category. However, as a zoning measure, the level of that activity is not evaluated. Without any data, the reversal of the agricultural category and the commercial and industrial category should be considered.

Ecological Receptors

Of the seven features of altered ecosystem function cited by Neuhold and Ruggerio (1976) –primary productivity, secondary productivity and growth, reproduction and development (including sex ratios and age and size class structure of the population), nutrient cycling, community diversity and structure, keystone species (a species that determines the character or nature of an ecosystem), and valuable species (including, but not limited to, endangered species and their habitats)–only the last-named feature is designated as a receptor, and then only to the extent that it is officially recognized or designated as such by other agencies.

That habitat once suitable for an endangered or threatened species has been degraded by toxic releases appears to be of no consequence to the scoring exercise. The restoration of the species, or even a reduced rate of degradation, is apparently not meritorious enough. Should a habitat be scored for its overall biotic potential for supporting threatened and endangered species?

The size of the potential critical environment is scored higher for extent than for the actually more critical "patchiness" of such environments. For some species with large territories (e.g., mink and bald eagle), extirpation has proceeded in spite of opportunity for immigration. For nonmobile species and many rare plants, adequate habitats might be scattered, yet very vulnerable to disturbance by released chemicals. As stated in the scoring box (DOD, 1991b, p. 80), an area of irrigated farm land larger than 100 acres would receive a higher score than an area of patchy natural wetlands and forest smaller than 100 acres, even though the latter might have greater biotic diversity (including species classified as lower than the highest category of endangered). Should the score take cognizance of this critical ecological parameter?

The aesthetic and recreational values of minimally managed areas are specifically excluded from scoring. For example, there may be merit in including wilderness as a scorable attribute. One means might be, as suggested above, to develop an additional factor based on the inversion of the land use and/or zoning factor in the DPM.

Many military reservations create valuable habitat, particularly in buffer zones or other areas held in nonuse for long periods. That adds a complex feature to the scoring exercise. If the land were to be decontaminated and land use altered to civilian status, such as residential development, the ecological value of the area would be degraded, so assigning a high ecological score to the site (and consequently a high priority for remediation) might have the long-term result of ecological degradation. Until a transfer to civilian status occurs, protection of such habitat within the boundaries of the installation would be important because it is unlikely that such a habitat is specifically recognized or federally designated as critical. For example, Rocky Mountain Arsenal probably has some 40 acres of undisturbed tall grass prairie and large areas of land that have been out of cultivation for over 50 years. Clearly, the ecological score should express the ecological value or potential value of on-base lands.

Terrestrial habitats of threatened species or of species not otherwise designated as endangered score 0, an unusual score for an air and soil receptor. Because the air and soil pathways are generally applied only to plant species, it would not be anticipated that a critical coastal environment less than 1 mile away would score 13 but a critical freshwater environment the same distance away would score only 6. The above treatment is unusual given that there is no specific information in the literature implying that the sensitivity of a coastal environment is twice as great as a freshwater environment with respect to these pathways.

DOD's Priority Setting

SOCIOECONOMIC ISSUES

Socioeconomic Impacts of Site Remediation

The DPM model is deficient in its treatment of the important social and economic dimensions of receptor impacts. It is not clear whether DOD addresses these limitations through considerations external to DPM. The major shortcomings are

- the arbitrariness of the choice of categories for population densities, distances from sites, and land-use characteristics of the pathways;
- the lack of justification for attaching greater weight, and hence greater importance, to human impacts than ecological ones by a factor equal to the square root of 5; and
- the absence of attention to economic and social impacts of contaminated sites and to the cost-effectiveness of risk reduction.

The population and distance parameters built into the DPM tend to underweight very sparsely and very densely settled areas. Too much weight is given to moderate density sites. By using arbitrary classifications of land use (completely remote, agricultural, commercial-industrial, and residential), the DPM model ignores a key potential decision-making variable: the value of land and existing structures if a site were remediated. In other words, important opportunity costs are not considered.

The scheme of human and ecological weighting, if used too rigidly, influences the final ranking in favor of funding cleanup of human settlements. For example, impact on the population of a logging village would be scored higher than impact on the forest, although a population might depend on the forest for jobs. The weighting might also fail at the other extreme: the importance of protecting population could be underweighted by use of a ratio of only the square root of 5 to 1 in the case of a very large city.

Unless the current and potential value of property is estimated, it will not be possible to choose the most cost-effective method of cleanup. An example is choosing to clean up a site because it is inexpensive to remediate, rather than cleaning up one whose remediation would be more expensive but have great economic value to the community.

Socioeconomic Information in the DPM

The DPM process requires scoring for the number of people living and working within 1,500 feet, 2 miles, and 4 miles from the site and the installation and scoring for land use and zoning within 2 miles. The DPM rationale does not explain or justify why 1,500 feet is relevant to contaminated waters, 2 miles is relevant to other water exposures, and 4 miles is relevant to air pollution. If the population data were to be used for air emissions, one would have expected the data could be disaggregated into four to 16 cardinal compass zones and the potential exposure to the population estimated with a wind rose.

The DPM FY 90 Scoring Quality Assurance Program Report (DOD, 1990a) indicates that most of the population data were "estimated using personal knowledge of the scorer." Although the report noted that it applied to a model earlier than Version 3.0 (DOD, 1990b), it added that the scorers had to estimate the location of the population and that data sources were documented in only 12% of cases. In addition, as stated in the DPM Manual (DOD, 1991b), "several of the scorers entered smaller populations for larger areas." In the DPM, the land-use and zoning data are divided into four categories: completely remote, agricultural, commercial and industrial, and residential. The thinking behind the weights attached to the four categories is not explained. Nor is

DOD's Priority Setting

there an explanation of the dichotomy between distance to commercial-industrial land and distance to national or state parks, forests, wildlife reserves, and residential areas.

DOD sites (more so than civilian sites) tend to be on large tracts of land wholly controlled by military installations. Yet, it may be a mistake to assume that social and economic impacts are irrelevant for employees and residents on those installations.

SCORING METHODOLOGY AND AGGREGATION

Overview

The mathematical operations of the DPM are straightforward; they involve substituting values into formulas and, in some cases following logical branching. Although the mathematics is clear, the choices of particular operations for combining quantities appear to be somewhat arbitrary. For example, some scores are combined by a root-mean-square operation, and the result is an implied square-root-of-5 weighting of human-health subscores versus ecological subscores (DOD, 1991b). Also, the transformation of quantities via multipliers from continuous values to ordinal values or from descriptive phrases to numerical values appears to be arbitrary and without explanation (DOD, 1991b). The aggregation of several risk elements or other criteria into a single priority-setting index involves subjective values, as well as scientific judgments. But there are systematic and tested techniques for aggregating scores like this (NRC, 1992).

Final Aggregation

In the DPM, a final site score is obtained by aggregation of the eight subscores for pathway-receptor combinations (DOD, 1991b). The given aggregation formula uses a root-mean-square combination of the subscores as is also done in EPA's Hazard Ranking System (see Chapter 4). Apparently that choice was made because, compared with the arithmetic mean, it permits "individually higher subscores to have a greater effect on the overall score." The rationale for the above approach is unclear. For example, it would be even more true for roots of higher powers, such as the root-mean-cube. The reasons why root-mean-square is common in statistical and engineering calculations seem irrelevant here; perhaps a larger exponent would yield a spread of site scores more helpful for allocating resources.

The given aggregate formula implicitly treats the scores to be combined as independent. The assumption is that there are neither significant interactions (e.g., a health insult through one pathway exacerbating the health impact of some intake via another pathway) nor significant double counting. In the present version of the DPM, volatiles and dust are treated separately for air and soil transport. But for human and ecological receptors, only the larger of the two resulting scores is entered into the aggregation, the other being dropped entirely (DOD, 1991b). This exemplifies the failure of the DPM approach to aggregate the effect of two contaminant release sources that could lead to a higher potential threat than the individual sources alone.

Clear justification for various multipliers in scoring algorithms is not provided. Thus, it appears that the individual pathway scores and their aggregation cannot be analyzed along a systematic theoretical basis that might enable one to check the rationale in the scoring methodology and to propose revisions. Of particular concern is the transformation of continuous or cardinal datum (e.g., distance to nearest surface water) to an ordinal score (e.g., 0, 1, 2, 3) for which the rationales for the threshold levels used in the cardinal-to-ordinal transformation are not given. As noted in Chapters 3 and 4, the lack of a rigorous, testable basis for determining such factor scores is a problem common to structured-value models such as the HRS and DPM.

VALIDATION

Model validation encompasses the soundness and accuracy of the model as a means of establishing priorities for remedial action, as well as the mathematical and numerical aspects of the model computer code. The DPM has not been validated, even though validation is recognized as critical in the development and application of models for use in policy and regulatory decisions (Naylor and Finger, 1967; Chapra and Reckhow, 1983; Reckhow et al., 1990; Arula, 1987; Shaeffer, 1980; ASTM, 1984; EPA, 1989b; NRC, 1990b). A broad validation can be performed fully only in the context of the intended model use; that is, does the model give good advice in establishing priorities for remedial action? The direct output of the model, the DPM score, is intended to provide a measure of relative site risk or threat, which is intended to be used in the setting of priorities for resource allocation. Validation efforts need to address not only the relative measure of risk provided by the output scores, but also the quality of the rankings that result. Possibilities for the latter might involve classifying the sites simply into groups (e.g., high, medium, or low priority for remediation) as a result of the scoring or making a finer detailed ranking of the site scores.

An appropriate validation study–comparing model results with what they should be–would involve perhaps 10 to 30 sites and the comparison of scores and rankings from the DPM with those from another approach, assumed a priori more likely to yield the right answer. Ideally, the sites used would be authentic ones, perhaps including some already selected for rapid cleanup outside the DPM framework, but they could also include hypothetically designed sites. The purported right answers could be based on the results of independent risk analyses performed according to well-established procedures or on the judgment of experts who can give those sites intense consideration. Similar comparisons have been made for the EPA Hazard Ranking System models (Doty and Travis, 1990). For authentic sites that have advanced to later stages of investigation, the right answers could be based on experience involving the manifestation of a hazard or the benefits realized through remediation. A record of the validation study's procedures and results should become part of the DPM's documentation.

SENSITIVITY AND UNCERTAINTY ANALYSES

An important step in evaluating the performance and reliability of priority-setting models is to determine through sensitivity and uncertainty analyses the magnitude of uncertainties in the model site scores and the implications of the uncertainties for site ranking.

Detailed sensitivity and uncertainty analyses are yet to be performed on the direct model output (the DPM score) and on the resulting site rankings or priorities. The latter can be examined by determining how uncertainties in DPM scores affect an overall ranking and inclusion on the short list of sites identified for highest-priority consideration. Uncertainties in model output can be derived from assumed uncertainties in model inputs (or structure) or evaluated directly by analyzing how site scores vary among different analysts.

To illustrate the potential impacts of uncertainties in the DPM model results on ranking, a preliminary analysis was recently conducted (NRC, 1992) based on the available set of FY 1991 DPM scores (M. Read, DOD, pers. comm., July 1991). In this analysis, the uncertainty in the composition of the list of FY 1991 sites that had the 50 highest scores was evaluated. The changes in scores that resulted from the FY 1991 DOD quality-assurance (QA) review were used to scale the uncertainty in site scores.

The uncertainty assumed for the scores was based on the magnitude of the changes that occurred for the 50 sites for FY 1991 that underwent the QA correction. This uncertainty was superimposed on the full set of 284 sites scored in the DPM by DOD in FY 1991. The DPM scores for FY 1991 ranged from 1 to 64, with the distribution shown in Figure 5-4. Because all the scores that will ultimately be used for DOD priority setting will already have undergone QA, an additional analysis was performed on the assumption that the uncertainty in site scores is only one-fourth

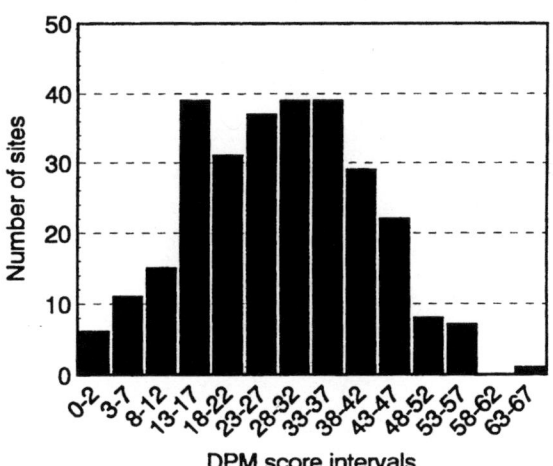

FIGURE 5-4. Grouped frequency distribution of DPM scores for 284 sites in Fiscal Year 1991. Of 284 sites scored, 15 have DPM scores in an interval from 8 to 12. Highest-ranked site is Rocky Mountain Arsenal, which has a score of 64. Two sites, Riverbank Army Ammunition Plant E/P Ponds and Richards Gebaur Hazardous Waste Drum Storage Site 923, have a low DPM score of 1.

that reflected in the QA corrections. Although this change in assumption did significantly reduce the variation in rankings, in both cases the composition of the group of sites with the highest rankings proved subject to considerable variation.

To implement the analysis, fifty simulations were performed by adding random-error values with the properly scaled variance to the FY 1991 DPM scores. The modified scores and overall ratings were determined for the 284 sites. The results of the analysis are summarized in Figure 5-5, which presents the uncertainty in the composition of the top-50 list for the two cases. It shows the probability, or fraction, of simulations in which each of the 284 FY 1991 sites is included among the top-50 ranked sites. As shown in Figure 5-5a, the variation in the full-QA-uncertainty case is quite large. Sites with baseline scores as low as 12 have a nonzero chance of being included among the top-50 sites (corresponding to a least one simulation in which the site was among the top-50 scores). Fully 219 of the 284 sites were included among the top 50 sites in at least one simulation. Furthermore, the top baseline site with a score of 64 was included among the top-50 sites only 84% of the time (corresponding to 42 of the 50 simulations). If the magnitude of site-score uncertainty is in fact comparable with that reflected in the FY 1991 QA modifications, then decisions based on the DPM ranking (e.g., to begin remediation with the top 50 sites) are subject to considerable uncertainty.

The variation in the top-50 list for the one-fourth-QA-uncertainty case is shown in Figure 5-5b. The impact of site-score uncertainty is shown to be greatly reduced. Sites with baseline scores above 50 are virtually assured of being on the list, whereas sites with baseline scores below 30 are virtually assured of being left off the list. The number of sites included in the top-50 list at least once is reduced to 107, compared with 219 in the full-QA-variability case. Site rankings and remediation decisions are thus more robust with this lower level of site-score uncertainty.

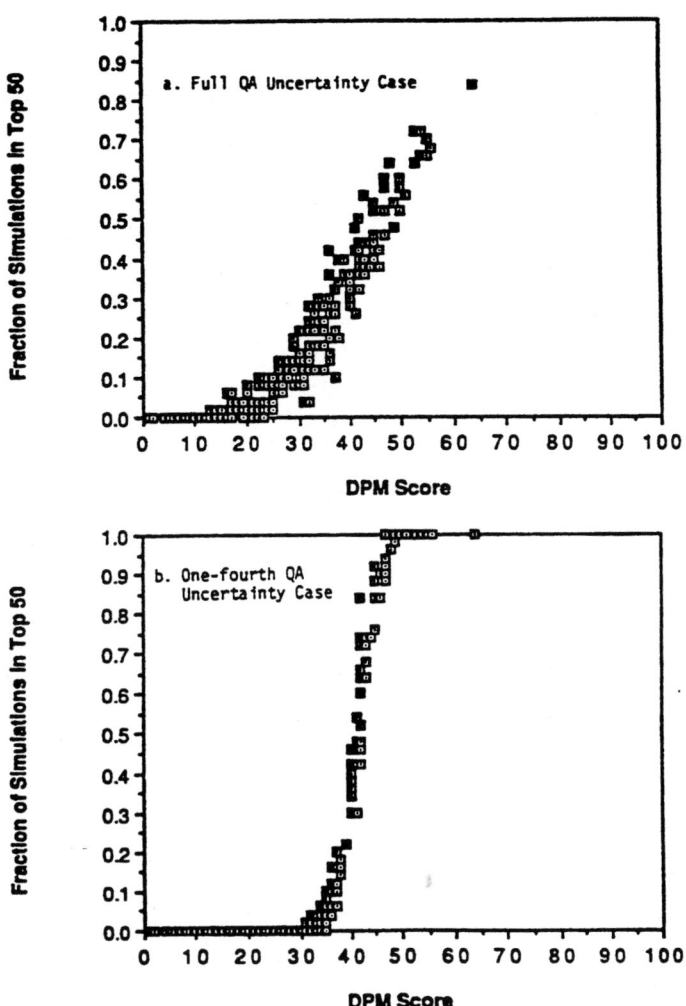

FIGURE 5-5 (a: top; b: bottom) Probability of inclusion among top-50 ranked sites as function of final Fiscal Year 1991 DPM score. Results based on 50 Monte Carlo simulations.

The analysis presented in Figure 5-5 can be extended to focus attention on sites that warrant additional study and effort. For example, in the case of one-fourth QA uncertainty, sites with scores above 50 are so likely to belong to the top-50 list, and sites with scores below 30 so unlikely to belong, that they require little additional study to reduce the uncertainty in their scores. The composition of the top-50 list is sensitive only to uncertainties in sites with scores of 30 to 50. It is those transition sites that should be targeted for further study to reduce their site-score uncertainty. The uncertainty analysis thus provides a mechanism for focusing further data collection and study efforts.

This analysis illustrates the kinds of sensitivity and uncertainty evaluation that could be performed. The uncertainty in site scores is shown to have a considerable impact on the composition of the top-50 list, although this impact is sensitive to the magnitude of the assumed site-score uncertainty. The analysis demonstrates that the uncertainty in DPM scores could potentially limit the use of the model for setting priorities among sites for remediation. It is essential that uncertainty evaluation of this type be performed for priority-setting models in the context of their intended use.

SUMMARY

The DPM is structured as a user-friendly model, and QA/QC approaches are used in its application. The DPM approach of using the product of pathway potential, hazard, and receptor is one reasonable approach to defining an overall site score. A detailed review of the DPM, however, reveals that the some of the transport and fate algorithms, toxicologic and exposure assumptions, and methods embedded in the DPM have weak theoretical foundations. For example, the fate algorithms used for the surface water, groundwater, and air and soil characteristics do not have an acceptable theoretical basis.

The pathway algorithms in the DPM use a summation formula to reflect the combined effect of the various pathway parameters, whereas theory suggests that a product of characteristics or a sum on a logrithmic scale would be the preferred approach to score the pathway potential.

The DPM does not appear to explicitly address social and economic impacts on site characterization, and it is not clear whether DOD addresses these issues through an evaluation process external to the DPM.

The DPM scoring scale is linear, and results of DPM site scores reveal that the score intervals for the FY 1990 and FY 1991 sites are small. Thus, based on the above results, it appears that the DPM may have a limited capability to discriminate between sites (NRC, 1992). Score compactness on the 0-100 score interval and factors in the pathway summation algorithms that mitigate against discrimination on pathway potential suggest that model algorithms should be restructured to produce a logarithmic scoring scale. Spreading out the numerical scores with alternate algorithms to combine scores, such as product algorithms (see NRC, 1992), might allow better discrimination between site scores. A simple sensitivity analysis for 50 highest DPM scoring sites (NRC, 1992) demonstrated that uncertainties can have large effects on the composition of the top-50 list.

To provide a summary evaluation of the tools used in DOD's priority-setting process, the criteria identified in Chapter 2 for effective model development and application and for the specific technical desired features are examined with primary focus on the DPM model. Given that the DPM was undergoing development when the committee performed its analysis, the comments given below should be viewed as committee recommendations of needs for such future efforts, rather than as evaluative of a completed product.

General Issues in DPM Model Development and Application

Defined Purpose: The DPM has a well-defined purpose within DOD's overall priority setting process. It is responsive to the stated DOD policy of giving site-cleanup priority to sites which present the greatest potential threat to human health and the environment.

Credibility and Acceptability: The extensive scientific peer review, public participation, and public comments that are needed for establishing credibility and acceptability of a model to be used in priority-setting have not yet been conducted with the DPM.

Appropriate Logic and Implementation of Mathematics: The calculation methods used in DPM are fairly straightforward. However, the logic of the choices made for particular operations and for combining quantities appears somewhat arbitrary. The basis for combining and weighting is not clear from the information which was provided to the committee.

Model Documentation: The documentation for the model was limited at the time the committee performed its analysis. That which was available to the committee does not adequately explain why the model is designed as it is. More extensive documentation is needed to describe the whole modeling process, as well as its product. For example, many of DPM's features, instead of being self-evidently correct, appear to be choices among many possible options. Such choices call for explanation and support. Documentation is also needed for evaluation of whether some particular kind of risk is being quantified consistently and whether the model's default values were chosen to be consistent with some explicitly stated policy.

Model Validation: The DPM has not yet been validated. Complex models such as this need to be checked carefully to determine whether in fact they perform sufficiently well for their intended purpose. The question is whether or not the DPM gives good advice in establishing priorities for remedial action. This has not yet been adequately determined.

Model Sensitivity and Uncertainty Analysis: There are many uncertainties in the data collected from a site for use in priority setting models. The effect of such uncertainties in model outcome and the priorities that sites receive should be known. An adequate sensitivity and uncertainty analysis for the DPM model is needed.

Specific DPM Technical Features

Applicability to All Waste Sites: The DPM is broadly applicable to essentially all DOD sites for which the model might be used.

Allowance for Dynamic Tracking: The DPM has not been developed as a tool for dynamic tracking.

Discrimination between Immediate and Long-Term Risk: The purpose of the model is to address primarily the longer-term risks. Immediate risks will be addressed as a first priority of the DOD and will not be subjected to priority-setting through the DPM.

Inclusion of Cost Estimates of Remediation: The DPM does not consider cost issues or timing with respect to remediation.

Transparency: The DPM is a highly transparent model. The mathematical formulation used are well described, and the procedures used for weighing of health and environmental risks are readily apparent. Model transparency and simplicity in use are major advantages of this model.

User-Friendliness: The DPM scoring procedures are straightforward and are described in an easy-to-follow procedure.

Appropriate Security: It is not clear at the moment how security issues will be addressed.

DOD Priority-Setting Process

Unlike the EPA's HRS, the DPM is used later in the priority-setting process after a more detailed site characterization, representative of a remedial investigation and feasibility study (RI/FS), has been completed. The DPM is not intended for use in ranking of all contaminated sites at DOD facilities. Apparently, sites posing imminent threats from hazardous or toxic substances will receive top priority for cleanup, and will not become part of the DPM evaluation. In addition, DOD plans to place higher priority on cleanup of sites on DOD installations that are subject to closure. Up until recently, site cleanup has not been restricted by a lack of funds. However, with increasing number of sites with detailed characterization completed, competition for funds is expected to become evident soon. The DPM does not explicitly evaluate the social and economic effects often associated with hazardous waste sites. It is intended only to provide a relative ranking of sites based upon their relative threat to human health and the environment. The ranking provided by DPM is to be used "along with additional risk information and other factors such as regulatory requirements and program efficiencies" to establish cleanup priorities among the DOD sites. The process by which other factors will be considered in setting priorities is not known to the committee, thus the committee is unable to comment on how well the overall priority-setting process will address the several features and technical issues noted above.

6

DOE's Priority Setting

For over 45 years, the U.S. Department of Energy (DOE) and its predecessor agencies have operated a large number of laboratories, chemical processing and metal manufacturing plants, nuclear reactors, and testing grounds for the primary purpose of producing nuclear materials and weapons for national defense. As a result of these operations, DOE facilities are now faced with contamination problems associated with radioactive and hazardous chemical wastes and mixtures of those wastes. These environmental problems, which exist at nearly 100 facilities and sites in some 30 states and territories, must be evaluated in order to determine the appropriate remedial responses.

To help manage the environmental cleanup and waste management activities, DOE has proposed an Environmental Restoration Priority System (ERPS) to "explain, document, and defend its budget request in discussions with OMB," and "to allocate [appropriated funds] among field offices, programs, and installations." (DOE, 1991a). After completing its analysis, the committee learned that

DOE had decided not to use ERPS for the foreseeable future. Although it is not currently in use, ERPS provided the committee with the opportunity to assess a model designed to explicitly address social, economic, political impacts, and cost consideration in addition to health risk impacts.

In this chapter, the background and history of ERPS is discussed, and the logic of the model is described. Special attention is paid to its technical underpinnings, as well as to issues that must be addressed to ensure its successful implementation. Before making this presentation, the committee notes that the description, analysis, and critique of the ERPS model is not as thorough as for the DOD and EPA counterparts in Chapters 4 and 5. There are three reasons for this difference. First, the committee was asked by DOD to conduct a detailed evaluation of DOD's DPM. It was not asked to perform a detailed evaluation of ERPS. Hence, in carrying out the study, the committee obtained much more documentation for the DPM than ERPS. For example, the committee had a computerized version of DPM and had an opportunity to run the model. A similar opportunity was not available for ERPS.

Second, ERPS is less than 5 years old. DOE was the last of the three federal agencies to develop a formal model to assist in priority setting. Consequently, there was much less published material available about ERPS upon which to base our analysis and critique than, for instance, on EPA's HRS. The committee was able to understand EPA's logic and to study in detail the data and equations used. A similar opportunity was not available for ERPS.

Third, ERPS is more comprehensive than DPM and HRS by explicitly including social, economic, and political impacts; cost considerations; and uncertainty in the model. However, the data base supporting these and other components of this complex model appear to be limited, and a considerable number of subjective judgments have to be made to run the model. The model has numerous scales (e.g., 1 to 4, 1 to 5, 1 to 7) and other calculations requiring squaring and taking square roots that are untested against actual cases. Some of the values and equations make sense, but they are at best educated guesses. To its credit, DOE has made these judgments explicit to decision-makers by building them into the model. In other words, ERPS allows more of the decision-making process to be built into the model rather than relying on an external process that is not open to scrutiny.

Because of these three factors, this chapter is more descriptive and less analytical than the chapters on the DPM and HRS. Although the committee presented criteria to evaluate DPM, HRS, and ERPS in Chapter 2, it feels that it had insufficient information with which to make a credible evaluation of ERPS. Information on health risk can be provided to ERPS through various alternatives. One of them is DOE's Multimedia Environmental Pollutant Assessment System (MEPAS). This chapter and Chapter 8 provide some evaluative information on MEPAS.

BACKGROUND AND HISTORY

Until the mid-1980s, DOE and its predecessor agencies took the position that DOE was essentially self-regulating with respect to environmental issues under the Atomic Energy Act. This interpretation was overthrown by a lawsuit. DOE now recognizes that its environmental restoration program is governed primarily by three major environmental statutes: (1) the Comprehensive Environmental Response, Compensation, and Liability Act (CERCLA), as amended, (2) the Resource Conservation and Recovery Act (RCRA), as amended, and (3) the National Environmental Policy Act (NEPA). In order to centralize and coordinate the efforts required by these statutes, DOE created the Office of Environmental Restoration and Waste Management (EM) in

1989, whose basic goal is to bring both active and inactive DOE facilities into compliance with applicable local, state, and federal laws and regulations. EM now oversees all DOE waste management and cleanup activities, including

• Waste management operations: minimizing, treating, storing, and disposing of wastes generated by activities at active facilities. This can be characterized as the RCRA component of DOE activities.
• Environmental restoration: assessing, cleaning up, and closing inactive sites and surplus facilities. This can be characterized as the Superfund component of DOE activities.
• Technology development: managing and implementing research and development related to DOE waste disposal operations and cleanup.

The DOE priority-setting system discussed in this chapter was developed initially for use with the environmental restoration program only. Another system for DOE's waste management operations, known as the Resource Allocation Support System has gone through the early stages of development. However, DOE completely suspended work on this system in 1992 and the committee did not review this approach at all.

In 1989, EM published its first five-year plan for cleaning up DOE's nuclear-related waste sites and for bringing its operating facilities into compliance with current environmental laws and regulations with the intention to revise this plan every year. The 1989 plan included the following four priority categories that were applied to environmental restoration and waste operations.

Priority 1. Activities necessary to prevent near-term adverse impacts to workers, the public, or the environment.
Priority 2. Activities required to meet the terms of agreements (in place or in negotiation) between DOE and local, state, and

federal agencies. These agreements represent legal commitments to complete activities on the schedule agreed to by DOE. (See Chapter 1.)

Priority 3. Activities required for compliance with external regulations that were not included in Priority 1 or Priority 2.

Priority 4. Activities that are not required by regulation but would be desirable (DOE, 1990).

A categorization system based on these four priorities was recognized as an interim approach to establishing priorities for future environmental restoration activities.

In the June 1990 executive summary of the EM Five-Year Plan for Fiscal Years 1992-1996, DOE reported that in consultation with interested parties, it was developing a risk-based priority-setting system for its environmental restoration program. The goals, in part, were "to support DOE budget formulation and allocation; measure the relative priority of program elements against a comprehensive set of program objectives, and explicitly identify the tradeoff among objectives," among others (DOE, 1990). According to the report, the approach was intended to be "a formal analytical decision-aiding tool addressing health and safety risks as well as social, technical, economic, and policy issues" (DOE, 1990).

Work on ERPS by DOE did not begin until 1989. From the start, efforts were made to solicit the views of interested and affected parties, both public and private. Involved parties are the State and Tribal Governmental Working Group (STGWG) and the Priority System External Review Group (PSERG), as well as technical groups. The members of STGWG include representatives of states and tribes that have negotiated Environmental Compliance Agreements and Orders with DOE that contain explicit near- and long-term milestones and schedules for cleanup and other environmental activities. Along with representatives from states and tribes, PSERG also has representatives from national environmental groups and EPA. In addition, broader public reaction was sought by seeking comments through a *Federal Register* notice on September 6, 1991, and by holding other such as a national workshop.

The initial publication on priority-setting system development (DOE, 1990) was followed by a more complete description (DOE, 1991a). Through a *Federal Register* announcement on September 6, 1991, DOE publicly sought comments on the priority-setting system (*Federal Register*, 1991). Fourteen organizations and individuals submitted comments. DOE published these comments and its detailed responses to them in a document entitled "Priority System Comment–Response Document." DOE also published a summary of the issues raised and DOE's response to them in the *Federal Register* (1992b).

In addition to these open review opportunities, DOE asked that a special ad hoc peer review committee evaluate the April 1991 report. This Technical Review Group (TRG), often called the Parker Committee after its Chairman, Dr. Frank Parker, published a report presenting the results of their deliberations (Technical Review Group of the Department of Energy, 1991). The TRG concluded that a formal prioritization system is preferable to an informal system. This methodology represents the state of the art. However, it has major limitations in what it can accomplish even with perfect input. Even with its current limitations, the system can play an important role in ordering priorities, but it is inappropriate for determining the budget for environmental restoration."

THE ENVIRONMENTAL RESTORATION PRIORITY SYSTEM

One of the objectives that is used for setting priorities in ERPS is the reduction of health risks. In order to measure the achievement of this objective, the priority-setting system recommends that formal risk assessments of specific sites be used. If a formal risk assessment has not been conducted, then the use of Risk Information Systems (RIS), output is suggested as an alternative for estimating population and individual risk estimates and for estimating risk urgency and timing. RIS outputs are summary risk indicators generated using a computer model known as Multimedia Environmental Pollutant Assessment System (MEPAS).

Since health risks play an important role in the determination of the priorities generated by the ERPS, the use of the MEPAS model deserves further comment. The MEPAS model was originally developed in 1988-89, and is based on a method known as the Remedial Action Priority System (see Michel (1992) and references therein for more background).

The MEPAS computer model was developed to specifically focus only on potential adverse health impacts, not on ecological or other types of impacts. Although MEPAS is not widely known outside the DOE community, it uses many elements that are better known than the model itself. It incudes elements that deal with radioactive wastes, hazardous chemical wastes, and mixtures of both types of waste. MEPAS uses mathematical formulas to predict the transport of chemical and radioactive contaminants to and through air, soil, surface water, and groundwater to where humans can be directly or indirectly exposed. Final exposure paths include direct exposure (through inhalation, skin exposure, drinking of water, etc.) as well as indirect exposure (eating of crops or fish, bathing, swimming, etc.) The MEPAS model can be run on a relatively simple computer system. In addition to estimating pollutant exposure amounts, which can be difficult or expensive to obtain, MEPAS, like HRS and DPM, needs other standard and often easily available data, such as rainfall, population density, soil type, and river flow.

The end result of MEPAS is a pair of risk estimates. The first risk estimate provides a maximum individual risk that can be predicted for the conditions specified for carcinogens (radionuclides as well as carcinogenic chemicals) and noncarcinogens. This first risk estimate is prepared from estimates of exposures caused by inhaling air; drinking water or milk; eating fish, vegetables, meat or soil; and recreational activities in contaminated surface water.

The second risk estimate is an indicator of risks to population groups: the Hazard Potential Index (HPI). It reflects the first risk estimates for individuals, weighted by the total number of people exposed. In general, it is standard practice to develop both individual and population risks to allow a more complete understanding of any given situation and to permit better-informed decisions.

Role of ERPS

The ERPS represents an application of multiattribute utility theory to the problem of ranking a set of alternatives for site remediation. Details of this system are reviewed by briefly discussing its role in the overall process of determining priorities and funding levels for environmental restoration activities. The basic structure of the model is then explained and evaluated, and the details of its implementation are discussed.

According to DOE, ERPS was intended to be used in support of its 5-year plan on an annual basis. The 5-year plan is DOE's basic management planning and budgeting document for EM activities. Updated annually, this plan describes all of the EM activities that DOE is conducting or is planning to conduct for the next few years.

ERPS would in principal be useful in supporting three aspects of the 5-year plan. In FY 93, for example, the system could be used to assist in allocating the actual funds appropriated by Congress for environmental restoration activities, in the event that Congress does not appropriate all the funds needed to carry out the activities previously planned for the current year. For FY 94, the system could be used to help explain and defend the proposed budget and activities before Congress during its deliberations. Finally, for FY 95, the system could be used to help consolidate the requests for funding environmental restoration activities at DOE facilities.

This use of the ERPS is illustrated in Figure 6-1 for the DOE 1995 budget-planning process. This process would begin in 1992 when the FY 95 requests for funding are received at DOE headquarters from the field offices and end in late 1994 when appropriations for FY 95 are issued to the field based on actual congressional appropriations and an analysis of priorities determined by EM.

An Overview of ERPS

ERPS, a multiattribute utility model, is set up to allow the simultaneous evaluation of multiple objectives:

- Reduce health risks
- Reduce environmental risks
- Avoid adverse socioeconomic impacts
- Respond to regulations
- Reduce uncertainty
- Avoid unnecessary economic costs
- Achieve DOE cleanup policy milestones

In addition, the system is set up to allow a two-step screening

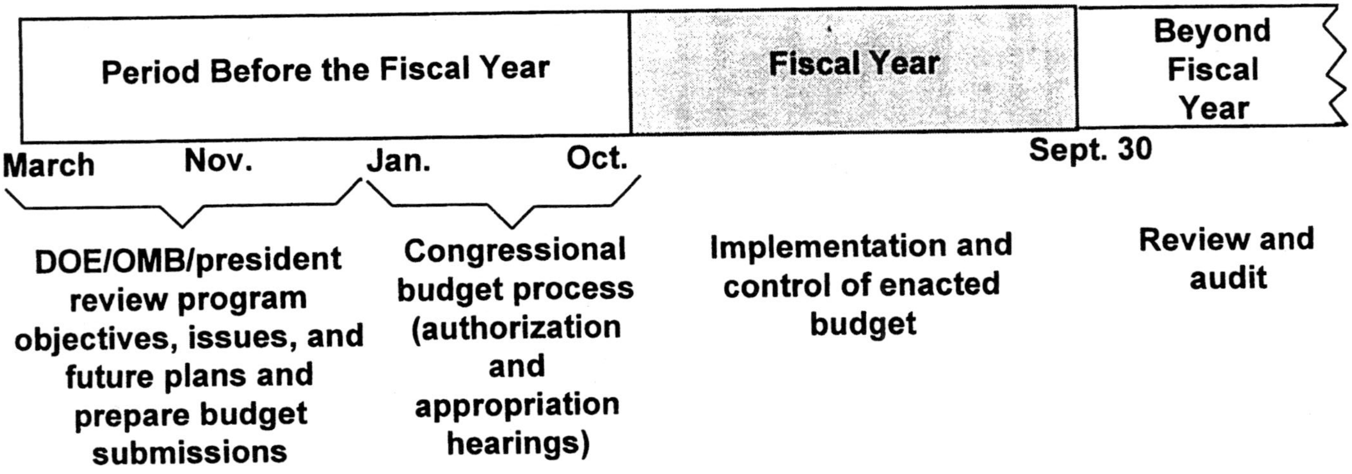

FIGURE 6-1 Overview of major steps in the federal DOE budget process. Source: DOE, 1992.

evaluation. These screening steps allow the determination of emergency activities and "time critical" activities. Emergency activities are self-explanatory. Time-critical activities are actions that are not emergencies, but need to be dealt with promptly; they include work to stabilize conditions that, if left untended, might deteriorate in ways that would create emergencies, and targeted study efforts to evaluate areas suspected of posing high, near-term risks. These two classes of activities are flagged for immediate funding. In effect, they are kicked out of the system at an early stage.

The rest of the system is then applied to other potential cleanup activities. The overall process is shown in Figure 6-1. The process incudes an initial phase carried out at the local level and a second phase at the national level. (The initial and second phases are not shown in Figure 6-1.)

At the local level, emergency and time-critical activities are identified and pulled out of the system at that point for immediate funding. All other possible activities are then identified and ranked (or scored) in accord with the seven objectives. Costs are also developed for each activity. Finally, activities are bundled into groupings called "budget cases." Each installation is charged to develop a small series of budget cases. These include a minimum budget case, which by definition will include time-critical activities and all other activities that meet the basic needs of the facility; other budget cases range upward to a maximum budget case, which includes all activities that the installation management believes that it could undertake in a given year.

The budget cases are then submitted to a central headquarters group where they are aggregated and analyzed for the entire set of environmental restoration operations. Both the estimated benefits (in the form of risks reduced or eliminated, for instance) and the estimated costs are covered in this analysis. The resulting analysis is then presented to DOE decision-makers as part of their consideration of alternative budget decisions.

Scientific Components of ERPS

The scientific evaluation of ERPS will focus on the methodology used for scoring the budget cases proposed by the field offices. An overview of the four steps required for multiattribute utility analysis is followed by a discussion of the details associated with their implementation in DOE's ERPS.

The Multiattribute Model

The ERPS is based on an implementation of multiattribute utility theory, an approach to ranking alternatives described in Chapter 3. From a scientific standpoint, multiattribute utility theory is a prescriptive approach to decision-making. It allows the orderly and simultaneous evaluation of multiple factors and objectives. To apply this technique in practice, three steps must be taken.

Step 1. Structure the decision problem. Structuring the decision problem includes identifying the alternatives (different actions that can be taken) and specifying objectives (goals of each different action with respect to DOE's seven objectives listed above), and attributes (what will be accomplished by each alternative). In ERPS, DOE field offices and installations play a major role in identifying and classifying budget-year activities and in defining budget cases (at least three) for evaluation using the multiattribute utility model.

The identification of the objectives is a critical step in the priority-setting model, since these same objectives are used by ERPS to compare budget cases from the same site, as well as to compare and to set priorities for budget cases from different sites. Therefore, the objectives have to be sufficiently general to make mentally different kinds of environmental restoration activities in different geographical, political, socioeconomic, and ecological environments. Yet, these same objectives have to be sufficiently detailed and unambiguously defined so that each budget case can be evaluated on each objective.

Although this list of objectives might be sufficiently general to cover a wide range of ERPS-related budget concerns, it is not sufficiently well defined to allow specific budget cases to be evaluated quantitatively against each objective. To render more operational this list of objectives for the purposes of ERPS it was necessary to create an objectives hierarchy with these seven broad statements at the top and with more detailed objectives further down. The lower level objectives essentially define the higher level objectives in more detail, and can be thought of as a means to their ends. The objective hierarchy for ERPS is shown in Figure 6-2.

For each of the lowest-level objectives in the hierarchy, an attribute must be identified that can be used to measure the degree to which the objective is achieved. For some lower-level objectives, such as cost, the appropriate attribute would obviously be in dollar amounts. For other objectives, especially those that relate to concepts, such as community concerns that are more subjective, it might be necessary to construct a metric attribute using a categorical scale (e.g., 1 = major impact, 2 = some impact, 3 = little impact).

Step 2. Assess the possible impacts of different alternatives. In this step, DOE attempts to measure the impact of each alternative on each attribute. In ERPS this step is carried out by "scoring" the budget cases on each of the criteria. This is an important step, and the scores for each of the attributes in ERPS are based on a different set of instructions. For each attribute, these instructions explain how the impact of a budget case is to be measured or estimated. In some instances, these impacts are measured on a natural numerical scale, such as dollars, while in other cases the

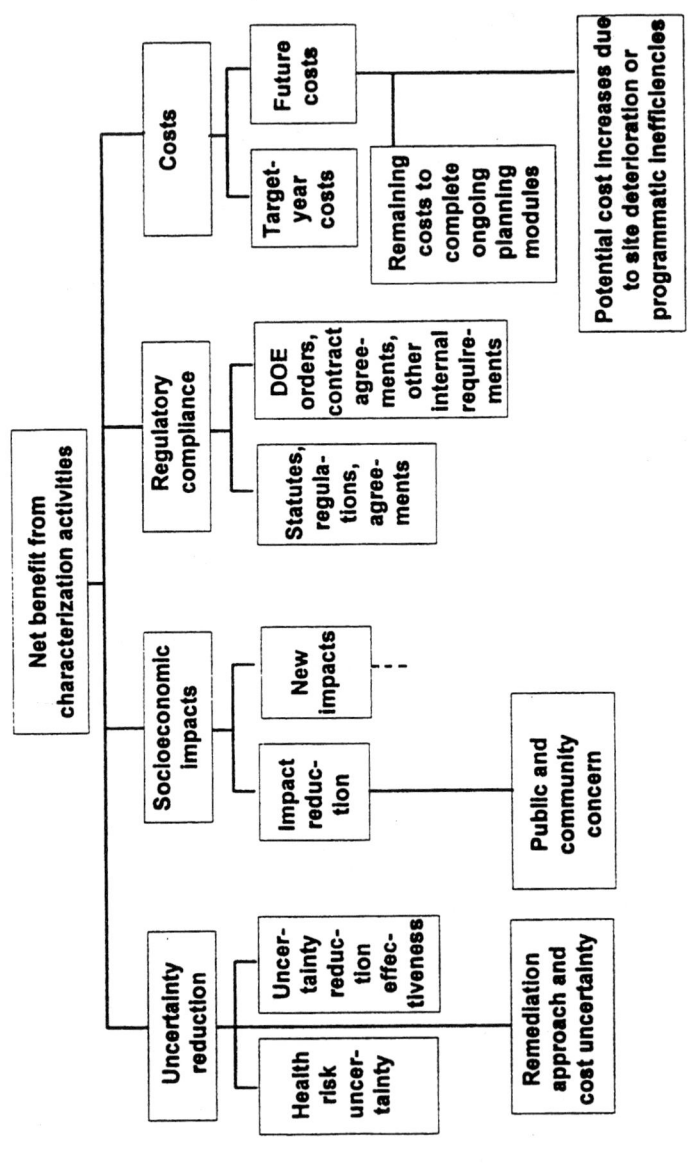

FIGURE 6-2 Criteria for evaluating characterization activities, arranged in a hierarchy. Dashed line under "new impacts" indicates criteria structure similar to that shown under "impact reduction." Source: DOE, 1992.

scales are verbal and require subjective judgments. These constructed scales allow scores that range from 1 to 7. For most of the attributes, scores close to 1 indicate favorable conditions and scores close to 7 indicate very unfavorable conditions. Further detailed discussion of the objectives and the instructions for scoring the attributes associated with them is presented in the next section.

Step 3. Determine preferences (values) for the decision-maker. The objective of Step 3 is to translate the scores assigned to each budget case on each alternative into a single number that can be used to rank the budget cases according to the preferences of the decision-maker(s). In the case of multiple attributes, this is accomplished by defining and assessing a multiattribute utility function. This process has four parts.

First, the functional form of the multiattribute utility function must be determined. Intuitively, the objective of this step is to define single-attribute utility functions over each of the individual attributes and then combine them mathematically to calculate an overall utility number for each alternative. In general, the single-attribute utility functions on the attributes could be combined in a variety of ways; for example, they could be added, multiplied, or combined using a complex polynomial form. However, if certain conditions of independence can be justified, then the form of the multiattribute utility function can be simplified to the additive case. In ERPS, the multiattribute utility model is additive at the level of the objectives.

Second, individual utility functions must be defined for the attributes and the objectives. These utility functions translate the scores on the attributes into measures of value or worth. The individual utility functions proposed on each of the attributes for the ERPS are different and will be discussed in the context of each of the objectives.

Third, weights must be determined for each of the objectives and attributes in an additive multiattribute model. Intuitively, it is often appealing to interpret these weights as measures of the relative importance of each criterion. More rigorously, these weights represent the relative importance of changing each attribute (or objective) from its least desirable to its most desirable level.

Fourth, alternatives must be evaluated and compared. Once the alternatives have been scored on each of the attributes, and their corresponding utility function values have been calculated from the single attribute utility functions, it is a straightforward matter to multiply these utility values times their weights and to sum the results to determine an overall utility function score for each alternative. This approach is followed for each of the budget cases in ERPS to determine their overall rankings. In addition, it is appropriate and typical to perform a sensitivity analysis on the results to investigate how the rankings might be affected if scores, single attribute utility functions, or objective weights are changed over a reasonable range. That is, the analyst systematically changes one element (for example, the weights from 0.1 to 0.6) to determine how much that element must be changed to alter the rankings. The goal of the sensitivity analysis is to identify the most sensitive elements in the model.

Defining the Seven Objectives in ERPS

How the scoring and the utility function values for the various attributes are determined for each of the seven objectives is discussed below, followed by a detailed description of the multi-attribute model that combines these assessments.

Health Risks

ERPS uses three measures of health risks that are eventually combined into two attributes: population health risk and individual health risk. The population health risk measures "the future expected peak annual health effects" in terms of the probability of an incidence of "a major, adverse health consequence." For this purpose, a major adverse health consequence is defined, in part, as a premature fatality, severe neurotoxic effect, or disabling birth defect (DOE, 1991b). The individual health risk is defined as the "estimated lifetime probability of a maximally exposed individual experiencing a major health effect." Finally, the third measure of health risk incorporated into ERPS is an estimate of health risk urgency and timing, which is defined as an estimate of the earliest time at which either the population or individual risks will actually equal their estimated values.

The scoring for the population and individual health risks is based on an approach that requires several steps. First, it is necessary to make an estimate of the baseline population health risk and the baseline individual health risk. These baseline risks are estimates of the potential health impacts that would occur if no future actions to reduce risks are taken. To determine the baseline population health risk score (on a scale from 1 to 7) for a site or facility, four possible sources of information are suggested:

• A formal risk assessment that explicitly estimates the future peak annual health effects on a quantitative basis.

• The use of judgment based on a subjective evaluation of future scenarios that could identify how adverse health effects could occur.

• The use of the "undiscounted aggregate HPI score" from MEPAS (HPI means Hazard Potential Index. It is an estimate of the long-term average health effects for a population (Droppo et al., 1990).

• The use of the Hazard Ranking System score from the model developed for EPA (see Chapter 4).

Guidelines for using information from any one of these measures to determine the baseline population health risk score are given in the instructions for implementing ERPS.

An estimate of the baseline individual health risk is also required. According to the instructions for ERPS, this estimate should be based on quantitative risk assessments. If no quantitative risk assessments have been performed, then the assumption is made that the individual risk score should equal the population risk score. The scores of 1 through 7 correspond to the estimated lifetime probabilities of a maximally exposed individual experiencing a major health effect ranging from 10^{-1} to 10^{-7}.

To evaluate a specific budget case proposed by a field office, an estimate is made of the reduction in risk (either population or individual) that would result from the associated remediation activities. This requires two estimates. First, the fraction of the risks in the baseline case that would be addressed by the remediation activities is specified based on subjective judgment. Second, the fraction of the risks that are addressed that will be eliminated is identified, again based on subjective judgment. These two numbers are multiplied together to determine the percentage risk reduction that would be associated with the remediation activities. This percentage is then transformed into a "scoring decrement" in the 1 to 7 units that measured the baseline risk score using a scale provided in the instructions, or a simple mathematical formula. Finally, this scoring decrement is subtracted from the baseline risk score to determine the case risk score for the corresponding budget case.

In summary, the health effects (population and individual) of the budget cases are evaluated in terms of estimates of how much they reduce the risks associated with the baseline case. Each budget case is assigned two scores from 1 to 7, one for population risk and one for individual risk. Each of these scores can be translated into a corresponding probability of adverse health effects.

The next step is to use a single attribute utility function to translate these scores into measures of value. The transformation for each of these health risk measures is an exponential function, which makes each of these health measures linear in the numbers of anticipated health effects. In addition, the health risk urgency and timing score is used to determine over how many years the number of anticipated adverse health effects should be discounted using a constant discount rate. Therefore, this transformation means that the health risks are actually measured as estimates of the reductions in the discounted (for timing) number of adverse health effects that would result from the remediation efforts associated with the budget cases. (In discussing the process of discounting, Droppo et al. (1990) state that normally, a major impact that will occur tomorrow is of more concern than an equal impact that will occur 7,000 years from now.)

This procedure for estimating health risks could be criticized on several grounds. The most obvious is the lack of specific scientific data for most of the estimates that are required to determine the baseline risk scores and to determine the effectiveness of the proposed remediation activities.

The scoring and scaling functions eventually lead to the use of a preference function that is linear in terms of the number of anticipated adverse health impacts. For many criteria, including anticipated numbers of deaths, social decisions are often made that imply that these preferences are not linear, but instead exhibit increasing or decreasing marginal values. However, over the range of anticipated health impacts that is considered in this model, the assumption of linearity seems appropriate.

A more subtle issue is the use of population risks and individual risks to estimate the health risks of budget cases. The multiattribute utility model that eventually combines these measures is additive, which implies that these measures are independent from one another in their effect on the ranking of alternatives. The use of these two measures may be viewed as a simplistic attempt to address a complex issue in risk analysis and public choice, the issue of equity. The population risk measure is an estimate of the expected number of adverse health effects across the entire population that could be exposed to contamination. As a result, this measure could easily indicate a higher ranking for one site where there is a very low risk to any one individual within a large population that is exposed over another site where there is an extremely high risk for members of a small population. Some would argue that it is not equitable to allow the exposure of a small number of individuals to high risks even if the expected number of adverse health effects in the small population is lower than in the larger population. The attribute of individual risk allows a tradeoff to be made between the expected number of health impacts and the level of risk to the individual (or individuals) who receive maximum exposure. Although more sophisticated models of risk equity have been proposed for situations such as this, these two measures do allow consideration of this issue.

Another issue is that the notion of adverse health effects is not differentiated by degrees of severity or by the characteristics of the individuals that are exposed. Thus, a number of difficult considerations are not addressed by the approach in ERPS to estimate health risks, including the need to distinguish between the implications of an incidence of cancer in an 80-year-old individual versus a birth defect in a child. To the extent that these issues are important and could be differentiated at different sites, they would have to be considered outside of the priority-setting system because of the vast amount of data that would be needed to include the issues in the system.

Finally, the risk urgency and timing score is used to determine the number of years over which to discount the health effects not estimated to occur in the next 70 years (Droppo et al., 1990). If the discounting of health effects were not desired, this option is effectively eliminated from the model by setting the discount rate equal to zero.

Environmental Risks

The procedure with each budget case for evaluating the impacts on environmental risks is similar to the one used for health risks. First, a baseline environmental risk score is established for the site. Then the same procedure is followed for each budget case (using the same health risk worksheet) to determine estimates of the percentage of the environmental risks that would be addressed by the remedial activities associated with the case and the effectiveness of these activities. This determines a scoring decrement to be subtracted from the baseline score.

The baseline score for the environmental risk at a site is determined from two factors. The first is a measure of the sensitivity of the environmental resources at risk. The second is a measure of the environmental threat associated with the problems at the installation.

The score for sensitivity is determined by listing each major environmental resource and assigning it a score from 1 to 4, where 1 corresponds to a resource that is relatively less sensitive (e.g., a state-designated natural area) and 4 indicates a resource that is extremely sensitive (e.g., a national park). A worksheet is provided to aid in this scoring. The total environmental sensitivity score for a site is simply the sum of the scores associated with its individual resources. Next, the estimated level of threat is assigned a score from 1 to 5 depending on a judgment regarding the likelihood that the environment could actually be contaminated as a result of the installation's waste problems. Specific examples are given of situations that should be associated with each score. Finally, the baseline environmental score is determined by the product of the measure of the environmental sensitivity and the magnitude of the environmental threat. The resulting product is scaled to be between 1 and 7, the same scale for the scores assigned to the health risks.

The environmental score is assumed to be a logarithmic scale and is transformed by an exponential single attribute utility function to create a measure of value in the multiattribute utility function. From a scientific standpoint, this measure of environmental impact may be intuitively appealing, but it is a crude measure at best. This procedure illustrates a situation where an attempt has been made to simplify the implementation of the priority-setting system at the potential cost of some accuracy in the estimates. Direct estimates of the value of the remedial efforts associated with budget cases in alleviating environmental threats might be more accurate than these scores.

Socioeconomic Impacts

A score for the objective of minimizing negative socioeconomic impacts is derived from a simple model that combines the scores on three attributes. Once again, the scores are assigned from 1 to 7 for each attribute.

The first attribute related to the socioeconomic impact of a budget case is the level of public concern over the installation's problems. A score of 1 is used to indicate that the area residents demonstrate almost no knowledge or interest in the waste problems of the installation. A score of 7 indicates a very high level of concern, that might be characterized by frequent negative national newscasts and large-scale protest.

The second attribute is a measure of the impact of the budget case on the cultural and religious principles that are prevalent in the area. A score of 1 would indicate very low impact, while a score of 7 would be used to indicate major conflicts.

The third attribute is a measure of the economic or opportunity losses associated with the impact of the budget case on the surrounding community. Examples of potential losses are: 1) depressed property values in close proximity to the installation; 2) frustrated demands to use contaminated facilities, lands, or other resources; and 3) decreased recreational value. If these losses are anticipated as a result of the selection of a budget case, an estimate of the economic equivalent of these losses is required. A score of 1 is assigned to estimated economic losses of no significance, while a score of 7 is assigned to estimated losses on the order of $100 million total or $10 million annually.

Once a budget case has been scored on these three attributes, its overall socioeconomic score is calculated by a nonlinear mathematical equation involving these three attribute scores. Specifically, each of the three scores is squared, the squares are summed, and the square root of this total is taken and normalized to the range of 1 through 7. This equation was determined by a panel of experts convened for this purpose.

Finally, the score for each budget case is transformed by an exponential single attribute utility function into a linear scale that measures the value of these socioeconomic impacts. From a scientific point of view, this approach is crude and again represents a trade-off between technical rigor and the need to create a model that is relatively easy to implement. However, since the transformation of the three attribute scores into an overall score for the socioeconomic impact of the budget case was reviewed by a panel of experts, it may be a reasonable approximation to a valid measure for this important, but rather vague, objective.

Uncertainty Reduction

Perhaps one of the most innovative aspects of the ERPS is the inclusion of "uncertainty reduction" as an objective in the multi-attribute utility model that is used to rank the budget cases. In the early stages of environmental restoration work, a great deal of uncertainty might exist at an installation regarding the actual amounts of various radioactive and toxic wastes, the risks to health and to the environment associated with these wastes, and the costs of the actions that will be necessary to mitigate these problems. Therefore, it is necessary to undertake some activities that may be classified as characterization activities, whose sole purpose is to reduce some of these risks.

One of the advantages of a formal analysis of a problem using decision analysis techniques is that the value of eliminating some of the uncertainties associated with alternatives (budget cases) can be explicitly determined in economic terms. These ideas have been incorporated into the ERPS to determine a score for the objective of uncertainty reduction. If this objective were not included, then activities and budget cases that focus on characterizing the nature and significance of the hazards associated with potential problems at an installation would not be valued appropriately by the priority-setting system.

The details of the procedure used to determine a score for the reduction of uncertainty that may be achieved by a budget case are complex and require responses to a worksheet with seven different parts. Therefore, a summary of the logic behind this worksheet will be given rather than a description of its individual steps.

The value of uncertainty reduction at a specific installation depends on (1) the degree of the uncertainty associated with the risks at the installation; (2) the degree of the uncertainty associated with the appropriate remediation activities and therefore with their total costs; and (3) the effectiveness of the proposed characterization activities in reducing the uncertainties associated with (1) and (2). The first parts of the worksheet for estimating the uncertainty reduction associated with a budget case require a best-guess estimate of the current risks associated with the problems being characterized and an estimate of how much higher those risks could be. These estimates are based on the scores that were developed to measure population and individual health risks.

The next parts of the worksheet require estimates of the range of possible remediation costs associated with the same problems. High and low cost estimates are required to determine an implied probability distribution over costs. Finally, judgments are required to estimate the effectiveness of characterization activities in eliminating all or only percentages of the risks associated with risks and costs. From this information, a uncertainty reduction score is determined on a scale from 1 to 7. This scale too is transformed by an exponential function into a linear scale that assigns value to the uncertainty reduction activities.

Conceptually, the inclusion of the objective of uncertainty reduction in the ERPS is a sound and important feature of the model, eventhough estimating the score for this objective is complicated and requires several judgments that would be difficult to make.

Regulatory Responsiveness

The determination of a score for the objective of being responsive to all regulations requires judgments regarding three issues: (1) How likely is it that a regulatory violation will occur? (2) How serious will the violation be if it occurs? and (3) How soon might the violation occur? Worksheets are provided to assist in making these judgments for the baseline situation at the facility and for the different budget cases.

The likelihood that a budget case will result in an allegation of a regulatory violation is estimated on a scale from 1 to 5, where a 1 indicates that the budget case will keep the facility on track with regard to regulations and a 5 indicates that it is almost certain that one or more violations will occur. In a similar manner, the seriousness of a violation, if it should occur, is scored on a range from 1 to 5, with a 1 corresponding to a violation of minor significance (e.g., a slippage in meeting a nonlegally binding or a nonenforceable agreement) and a 5 corresponding to a violation of major significance (e.g., numerous enforceable obligations will be missed, significant monetary penalties will be assessed, and DOE will be charged with knowingly and recklessly endangering public health, welfare, or the environment). Finally, the time at which the regulatory problem is likely to occur is scored on a 5-point scale.

These three scores are then combined into a total regulatory score scaled from 1 to 7 using a mathematical equation. Essentially, this equation corresponds to multiplying the seriousness of the violation by its likelihood, and discounting the result at 25% per year of delay. An exponential function is then used to translate this score into a measure of value for the multiattribute utility function.

Ability to Meet DOE Milestones

This objective is measured in a straightforward manner; it is not included as an objective in the multiattribute utility function. Very simply, if a budget case does not include sufficiently effective activities to permit achievement of DOE long-term policy milestones (e.g., the 30-year cleanup goal for DOE sites), then it is assigned a score of 1. Any case with a score of 1 must include a description of the specific milestone that is threatened.

Costs

Costs associated with the budget cases must also be estimated. Although the estimates of these costs may be difficult in practice, the requirements for determining them are straightforward.

First, the costs associated with the target budget year are estimated. Second, the remaining costs that are needed to complete multiyear remediation or characterization studies that are initiated or continued by the activities within the budget case are estimated. Third, any impacts on future costs are estimated. These costs could be, for example, the increased costs when a particular budget case forces a delay in remediation activities that would alleviate an existing, deteriorating problem.

Aggregation of Measures of Value

The steps described above lead to scores (typically from 1 to 7) and then to measures of value through the transformation of these scores with a single attribute utility function (typically an exponential transformation). These measures of value are then aggregated using the following multiattribute utility model:

$$\text{Utility of Benefits} = W_{pr}U_{pr}(S_{pr},S_u) + W_{ir}U_{ir}(S_{ir},S_u) \\ + W_{env}U_{env}(S_{env}) + W_{soc}U_{soc}(S_{soc}) \\ + W_{ur}U_{ur}(S_{ur}) + W_{rr}U_{rr}(S_{rr})$$

where the W's are weights on the objectives, the U's are the single attribute utility functions, and the S's are the scores on the objectives. The subscripts refer to the following objectives:

pr = population risks,
ir = individual risks,
u = risk urgency,
env = environmental risks,
soc = socioeconomic risks,
ur = uncertainty reduction, and
rr = regulatory responsiveness.

Finally, the net utility of each budget case is calculated from the expression

Net Utility = Utility of Benefits - $W_{rc}U_{rc}(RC) - W_{fc}U_{fc}(FC)$

where rc = remaining costs and FC = future costs.

The additive form of this multiattribute utility function implies the assumption of certain independence conditions regarding the preferences of the decision-makers for the different objectives. Technically, these independence conditions are known as preference independence and difference independence (for a discussion, see von Winterfeldt and Edwards, 1986). Informally, speaking, these conditions are satisfied if the preferences of a decision-maker regarding the performance of an alternative on one objective is not affected by the level of its performance on another objective. It seems likely that this technical condition might be violated for the objectives of minimizing population health risks and individual health risks. However, the practical implications of this violation, if it exists, should not alter significantly the rankings produced by ERPS.

The final issue to be considered is the selection of the weights (the Ws) in the multiattribute model. As discussed earlier, these weights should reflect the relative importance of changes from the least desirable to the most desirable levels of performance on each of the objectives.

In ERPS, the combination of logarithmic scores (the Ss) and exponential single attribute utility functions (the Us) has the implication that each of the objectives is measured on a linear scale. In this special case, it is also possible to interpret these weights as a trade-off ratio. In addition, because two of the objectives are measured in dollars, these trade-off ratios can be expressed in terms of the number of dollars that would be paid to gain a desirable unit increase in each of the other objectives.

In a pilot test of this model in 1990, the values for the weights that were used are shown in Table 6-1, along with their corresponding interpretations in terms of dollars. In this pilot test, the largest weight (36%) was placed on reducing health risks. An important issue for the final implementation of this model is the process that will be used to determine these weights, since the rankings of the budget cases will be very sensitive to this choice.

TABLE 6-1. Current weights for DOE ERPS.

Factor	Weight	Basis
Health risk	36%	$5M/health effect Individual weight = 1/5 of population weight
Uncertainty reduction	32%	Implied by weight on health and dollar tradeoffs
Environmental risk	13%	$400M to eliminate a "7"
Socioeconomic impact	9.5%	$300M to eliminate a "7"
Regulatory compliance	9.5%	$300M to eliminate a "7"

Other: Risk urgency/timing = 5% discount rate.
Remaining and future costs = 10% discount rate.

Source: T. Longo, DOE, unpulished data presented to the committee, April 10, 1991.

SUMMARY EVALUATION OF DOE'S PRIORITY SETTING FOR HAZARDOUS WASTE SITES

The DOE priority-setting process was less well developed during the period of the committee's activities and so it did not receive as much evaluation as EPA's HRS and DOD's DPM. Like those mathematical models, DOE's ERPS model was developed as an aid in DOE's overall priority-setting process. The ERPS model is much more comprehensive than that of the other agencies and considers social, economic, and political impacts as well as risk to human health and the environment. Although ERPS covers a much broader scope of considerations, it nevertheless should address the criteria identified in Chapter 2 for effective model development and application and should contain certain specific desired technical features. The following is a summary of the evaluation of these criteria with respect to ERPS to the extent that the committee could do so.

General Issues in ERPS Model Development and Application

Defined Purpose: The purpose of ERPS appears to be defined sufficiently well. It is to serve as an aid in allocating funds in a given year, in developing funding proposals to Congress in subsequent years, and in projecting fund requirements in future years. The model would help set priorities for these needs based upon a variety of social, economic, and political considerations, as well as impacts on human health and the environment.

Credibility and Acceptability: The committee could not evaluate this aspect sufficiently at this time. General acceptability of the model would be obtained through open evaluations by interested parties, good documentation, appropriate demonstration of the impact of model sensitivity and uncertainties, and good model validation.

Appropriate Logic and Implementation of Mathematics: Based on the committee's general assessment, ERPS is quite comprehensive.

Model Documentation: Adequate documentation of ERPS was not available at the time of the committee's deliberations.

Model Validation: ERPS is still under development, and a formal validation of the kind envisioned by the committee has not yet been conducted.

Model Sensitivity and Uncertainty Analysis: Inadequate information on these aspects was available to the committee for evaluation. One feature of the model is consideration of the value of reducing uncertainty through data collection and evaluation. Thus, the importance of uncertainty has been recognized by model developers. However, the model considers many different social, political, economic, environmental, and health issues in the development of rankings. In addition, many inputs to the model are based upon subjective judgements. It will be crucial in a broad and comprehensive model of this type to conduct a wide range of sensitivity analyses to help the model developers and others understand the impacts of uncertainties in data input on the resultant relative rankings of sites. The model will need modification in areas where small uncertainties lead to significant changes in budget case rankings.

Specific ERPS Technical Features

Applicability to All Waste Sites: ERPS is a comprehensive model that addresses both radioactive and nonradioactive hazardous wastes, and is it applicable to all DOE sites.

Allowance for Dynamic Tracking: Dynamic tracking is one of the important features for which ERPS was being developed. DOE planned to update model inputs on an annual basis and to consider in each year's funding allocations and requests the reductions of risk achieved at sites based on ongoing remediation efforts.

Discrimination Between Immediate- and Long-Term Risk: As with the other two agencies, DOE plans to address immediate risks outside of ERPS, and ERPS itself will be used to address longer-term problems.

Inclusion of Cost Estimates of Remediation: Cost estimation for remediation and associated risk reduction achieved are major features of ERPS.

Transparency: ERPS is quite transparent. The multiattribute utility theory that forms the basis for ERPS is a valuable tool for organizing a complex of information for review by others. The weighting factors used are readily apparent. However, the impact of changes in any given factor on the site-ranking outcome is not clear. For this reason, the sensitivity and uncertainty analyses indicated above will be needed.

User-Friendliness: ERPS allows input of subjective judgment in the scoring if the scorer does not have adequate scientific data available. This does lead to user-friendliness, but also could be subject to abuse and poor decision-making. The trade-off here between user-friendliness and sound ranking of sites needs consideration.

Appropriate Security: This aspect was not addressed by the committee.

DOE Priority Setting Process

Because ERPS was under development and has not been implemented for priority setting, the committee was unable to determine how well it would fit into the overall priority-setting process. ERPS is intended to be used to address cleanups that pose longer-term, rather than immediate, threats to public health and the environment. However, the committee also notes that DOE has entered into many agreements with EPA and various states that would override the relative evaluations of sites provided by ERPS. How DOE would address the conflicts likely to develop between such prior agreements and site rankings as developed by ERPS is not clear. A system such as ERPS could be an important tool in helping to address such conflicts by providing a more objective evaluation of desired remedial priorities. Thus, while the committee is not in a position to judge whether the ERPS would provide the objective ranking of sites for remediation that is needed, it nevertheless believes that a well-developed, documented, validated, and comprehensive model can help greatly in making sound decisions about what sites to remediate first, and the degree of cleanup that is desirable.

7

STATE PRIORITY SETTING

INTRODUCTION

Numerous states have encountered the need to develop ranking systems as an aid to setting priorities for the remediation of abandoned hazardous waste sites. States often have multiple statutes that provide authority for remediation of waste sites that are not covered under federal Superfund. The Environmental Law Institute (ELI, 1989, 1991) found that 24 states had their own priority-setting systems. Based on ELI's survey results, the committee asked a number of states to provide written descriptions of their systems. This chapter discusses the ranking systems of the states that provided descriptions. The approaches considered fall into three categories: systems similar to the EPA Hazard Ranking System (HRS) model; other explicit numeric systems leading to a site-specific score; and systems that categorize sites with the highest priority into three or more groups based on a narrative description of the severity of effects. This chapter examines several of the state-ranking models with respect to how they compare with each other and how well they achieve general objectives of a ranking system.

This chapter does not describe how the various state-ranking systems help to obtain a final priority for cleanup or the policy context of their application. In general, if observations at a site make it clear that a severe problem exists, a response is triggered even if that site did not receive a high numerical ranking. The documentation reviewed by the committee represented various stages of drafting and reformulation. Although some of the ranking methods considered in this chapter have not been made final, it is useful to examine them as examples of the different ideas on ranking methods that have arisen at various state environmental regulatory agencies.

STATES WITH RANKING SYSTEMS SIMILAR TO THE EPA HAZARD-RANKING SYSTEM

California, Ohio, Oregon, and Washington use indices of contamination severity—such as chemical toxicity, quantity, and mobility in the environment—that closely resemble those used in the EPA hazard ranking system (*Federal Register*, 1990; California Department of Toxic Substances Control, 1991; Oregon Department of Environmental Quality, 1991; EPA, 1992; Ohio Division of Emergency and Remedial Response, 1992; Washington State Department of Ecology, 1992). Scoring methods were provided by these states containing enough detail to allow a comparison of the scoring element values, routes of exposure, and algorithm structure.

State Priority Setting

Scoring Elements

For each of the four models, subscores are determined for up to four attributes of the contamination situation. Sites are ranked on a "worst first" basis with larger subscores reflecting greater concern for human health or environmental damage from the site. The following descriptions represent a composite of the four state models. The first attribute in the overall architecture is a subscore, called *release*, reflecting the strength of the evidence that contaminants are indeed present at the site in places and concentrations with the potential to migrate or cause a problem for direct on-site contact. This conclusion is supported, for example, by measurements of contaminant concentrations or observation of leaking containers.

For the most part, the second attribute—*substance characteristics*—scores the qualities of an individual contaminant in terms of its toxicity (human and environmental), mobility or water solubility, quantity, degree of persistence, and characterization of containment on site (e.g., landfill, above-ground container, or spill). This score describes the chemical contaminant in isolation from the topography and geology of the site.

The third attribute, *migration*, scores the same quality of chemical mobility as in substance characteristics, only from the perspective of the characteristics of the environment rather than the chemical. Parameters such as slope of the land, precipitation, and potential for flooding are rated so that greater potential for migration results in a greater score. California, in contrast to the other states, includes an exposure estimate in the migration section. Indices of exposure include distance to nearest structure and sensitive environment. This estimate would give higher scores to sites with close-by structures where the potential for human exposures would be greater.

The fourth attribute, *targets*, gives an indication of the receptors (human and environmental) that are in the vicinity of the waste site. The more and closer the potential receptors, the higher the score. Factors considered include: number of people living within 1 mile, presence of surface water bodies within 2 miles, population size served by wells, and the distance to the nearest well. The California model includes these sorts of considerations in both the migration and targets categories.

For each of these four attributes, a numeric score is determined by choosing from among site-specific numeric values. For example, migration potential is gauged by quantifying such parameters as amount of precipitation, slope of the land, and soil type. The degree of toxicity of a chemical is scored with parameters such as reference dose (RfD) for noncarcinogens and slope factor for carcinogens. For the most part, states have selected similar ranges of values to describe a particular parameter; however, significant differences exist in some cases. Table 7-1 shows a comparison of the states' middle values for several parameters pertinent to the groundwater pathway. Also shown for comparison are the corresponding values from the EPA HRS.

The most significant differences among the four states are in the scoring of the substance characteristics attribute. For example, the middle value of the quantity (in tons) of a contaminant is five times as large for Ohio as for California. This means that what would be considered a medium amount of waste in Ohio is considered a large amount in California. There are substantial differences between Oregon's treatment of the chemical toxicity indices for acute and chronic health effects as compared with those of the other three states. The LD_{50} leading to a medium score for Oregon is almost three times that for the other states. The differences in the RfDs is even more pronounced with Oregon's middle value 100 times higher than that of the other states. Therefore, the relative weight assigned to chemical toxicity compared to mobility and targets is lower in Oregon than in California, Ohio, and Washington.

STATE PRIORITY SETTING

Another point of comparison identified in Table 7-1, risk equivalent, is an expression of the relative weight given to the toxicity of carcinogens and noncarcinogens within a given ranking scheme. If two sites were to be compared, each having only one contaminant—a carcinogen at one site and a noncarcinogen at the other site—what would be the health risk at each site when both sites delivered the same dose of equally-weighted chemicals to a human receptor? In Oregon, for example, if exposure conditions were such that one were to receive exactly the RfD of a noncarcinogen at one site and the same dose of a carcinogen at another, the lifetime excess health risk from the site with the carcinogen would be 0.12. For California, Ohio, and Washington, the cancer risk at the RfD is 0.003. At a risk of 0.12, there is 40 times more of a given carcinogen than at a risk of 0.003. Therefore, Oregon treats carcinogens less stringently compared to noncarcinogens than the other states because at a specific dose (the RfD) a greater amount of a carcinogen is given the same rank as a noncarcinogen.

The values for the migration parameters are either identical or fairly close among the four states. There is approximately a factor of ten difference in the values of the target parameters between California and Oregon. For a medium score in the Oregon ranking system, more people need to be served by contaminated drinking water wells and more acres of land irrigated. Because the Oregon system has higher middle values for toxicology and target parameters, and approximately the same values for migration, the overall site rank would be relatively more influenced by contaminant migration than for the other states.

Routes of Exposure

To calculate a final score for a site, distinct routes of exposure to humans or other non-human environmental receptors are considered for up to the four site attributes (under consideration).

TABLE 7-1 Comparison of Selected Site-Specific Values Used in Scoring

Parameter	Middle Value of Range				HRS
	Calif.	Ohio	Ore.	Wash.	
Acute health effects, LD$_{50}$, mg/kg[a]	550	550	1,500	550	250
RfD, mg/kg-day[a]	0.00055	0.00055	0.05	0.00055	0.027
Slope factor, mg/kg-day[a]	5.5	5.5	2.5	5.5	0.5
Risk Equivalent[b]	0.003	0.003	0.12	0.003	0.008
MCL, µg/L[a]	55	55		55	
Quantity, tons[a]	50	260	110	125	
Coefficient of aqueous migration, K[a,c]		0.5	0.5	0.5	
Groundwater mobility and solubility, mg/L[a,c]	100	100	100	100	50
Precipitation, monthly, inches[c]	10		20	20	15
Depth to groundwater, feet[c]	100	100	100	100	138
Hydraulic conductivity, cm/sec[c]	10^{-5}	10^{-5}		10^{-5}	10^{-5}
Distance to drinking water well, miles[c,d]	0.5	0.5	1	0.5	0.75

Number of people served by drinking water wells[d]	100	1,200	3,000
Acres irrigated[d]	100	1,500	

[a]Substance characteristics.
[b]Does not appear in any ranking scheme, for analysis purposes only. It is an expression of the relative weight given to toxicity of carcinogens and noncarcinogens within a given ranking scheme.
[c]Migration.
[d]Targets.

Sources: Data from California Department of Toxic Substances Control, 1991; Ohio Division of Emergency and Remedial Response, 1992; Oregon Department of Environmental Quality, 1991; Washington State Department of Ecology, 1992; *Federal Register*, 1990.

The subscores for the attributes are combined to form a score for the route. Not all of the four states have considered the same routes for all of the attributes, but in most cases groundwater, surface water, air, and soil or direct contact have been evaluated. Table 7-2 shows the routes of exposure that have been considered by each state. In contrast to California, the other states consider surface water and air as different routes that depend on whether there is a human or environmental receptor. Ohio does not include a soil pathway or direct contact pathway because its model is not intended to deal with emergency conditions. Washington has a sediment pathway that is not present in the other state models.

TABLE 7-2 Routes of Exposure

California	Ohio	Oregon	Washington
Groundwater	Groundwater[a]	Groundwater[a]	Groundwater[a]
Surface water	Surface water[a]	Surface water[a]	Surface water[a]
	Surface water[b]	Surface water[b]	Surface water[b]
Air	Air[a]	Air[a]	Air[a]
	Air[b]	Air[b]	Air[b]
Soil		Direct contact[a]	Sediment[a]
			Sediment[b]

[a]Human.
[b]Environment.

Sources: Material from California Department of Toxic Substances Control, 1991; Ohio Division of Emergency and Remedial Response, 1992; Oregon Department of Environmental Quality, 1991; Washington State Department of Ecology, 1992.

Algorithm Structure

Although there is some similarity among the states in the selection of site-specific values (Table 7-1) and routes of exposure (Table 7-2), their mathematical operations for combining the resulting subscores are quite variable. Table 7-3 shows an example of the different methods used to combine values to arrive at the substance characteristics subscore.

TABLE 7-3 Substance Characteristics (SC) Subscore

California	Toxicity + Solubility + Waste quantity = SC
Ohio, Oregon, Washington	(Toxicity x Containment) + Waste quantity = SC

Sources: Material from California Department of Toxic Substances Control, 1991; Ohio Division of Emergency and Remedial Response, 1992; Oregon Department of Environmental Quality, 1991; Washington State Department of Ecology, 1992.

Solubility could be considered roughly equivalent to the extent of containment on site, which, together with toxicity and waste quantity, forms the basis for the substance characteristics score. California's model has only additions among the various attribute scores and routes. Toxicity is added to solubility, whereas the Ohio, Oregon, and Washington models multiply toxicity by containment before the result is added to waste quantity. All of the states ultimately arrive at subscores for all the routes, which are then combined into a total score for the site. California averages each of the four routes. Since no subscore for a route can exceed 100 points, the final score is between 0 and 100. Ohio uses the "root mean 4th power method." This is the fourth root of the mean of the route scores raised to the fourth power. Oregon takes the maximum route score, adds it to the mean of the other five routes, and adds ten bonus points if the site is in a sensitive environment. Ohio and Oregon have normalization procedures for generating final total scores between 0 and 100. Washington ranks subscores from 1 to 5 and combines ranks so that the highest pathway ranks are given more weight. The human health and environmental pathway ranks are combined in a matrix that weights human health more heavily. The final site score is between 1 and 5.

OTHER NUMERIC RANKING SYSTEMS

The Michigan ranking system provides a subscore for each of the following: environmental contamination, mobility, sensitive environment, population, toxicity, and waste quantity (see Chapter 1). These attributes closely parallel those considered in other state models and the EPA HRS. The method of scoring the attribute, however, is different from the other systems. For the most part, a series of narrative descriptions is associated with each attribute. For example, environmental contamination is scored by choosing from among 31 described conditions each of which has a point, as illustrated in the following example:

> One point shall be scored for surface water if a surface water body or wetland is located within 1/2 mile of the site, three points shall be scored for groundwater if a sheen is visible on an exposed groundwater surface, nine points shall be scored for surface water if the department of public health has issued a fish advisory for a water body and the cause of such an advisory can be attributed, in part, to the site (Environmental Response Division, 1990).

The use of variables, such as coefficient of aqueous migration, monthly precipitation, depth to groundwater, and hydraulic conductivity, is not found in this model. The final score is simply the sum of all the subscores.

STATE PRIORITY SETTING

STATES WITH RANKING BY BROAD CATEGORY

New York, Montana, and Missouri differentiate all sites into one of three categories of priority (Missouri Department of Natural Resources, 1991; Montana Department of Health and Environmental Sciences, 1991; New York State Department of Environmental Conservation, 1992). Each category is described by one to seven characteristics of the site that are predictive of its ultimate capacity to cause harm to human health or the environment. For example, New York, in its Category One (High Priority), requires a determination of "probable release to groundwater which is a drinking water supply." A Category Two site must demonstrate "minimal potential for release to groundwater that is a drinking water source," and Category Three "minimal potential for release to groundwater which is not a drinking water source." Other characteristics of each category include release to air, release to surface water, and effect upon a sensitive environment. Demonstrating that the criteria for any one of up to seven site characteristics per category are met is sufficient to rank the site in that category.

These ranking schemes provide an outline of what end points resulting from the toxicity, targets, and environmental fate of contaminants should be considered in determining the priority of cleanup. They differ from the HRS-like systems in that there is no mathematical combination of factors that lead to a score. Any one of a number of potential effects, if documented, could lead to a maximum score. This approach leaves the analyst much more flexibility in deciding which of the potential effects to pursue in more detail. For New York, this represents a change from a previously used numeric scoring system, but perhaps the state considers the added flexibility to more than compensate for the loss of quantitative information.

Discussion

The states have considerable collective experience with the problems of setting priorities for hazardous-waste site cleanup, and the need for and purpose of their ranking systems are clear. In general, the purpose and objectives of the states' ranking models are stated within the documentation accompanying the model. Often-stated objectives are that the models are intended to be scientifically defensible and easy to use, require minimal and inexpensive data, and provide results to help establish priorities for effective use of funds.

For many of the states considered, there is evidence of very thoughtful development of site ranking models such as parameters for location of fisheries, containment structures, population densities, and sources of drinking water. However, how the relationships between the model parameters were developed and what strategies were useful for combining the parameters is not always clear and often not documented, thus the models tend to lack credibility. There are many different scoring approaches among the states that use essentially the same type of data. It was not within the committee's scope of work to establish the reason for these differences or whether any of them offer a clear advantage over any others.

As was discussed in earlier chapters for the federal agencies' models, similar questions of appropriateness of the logic for combining various scores within the ranking methods apply to the state approaches. The documentation provided to the committee did not indicate the reasons—derived from first principles—for choosing the particular score combination approaches. The committee is unaware of comparisons of any state ranking system with other approaches such as risk assessment for the purpose of validation.

The extent and completeness of documentation varied considerably among the states' models. In some instances, another state was cited as the source of the major elements of an approach. For other states, very detailed data and explanations were provided in support of the range of choices for particular site variables. It appears that all the state models could be used readily by persons who have a minimum of formal scientific training and are provided with the necessary data.

States not using formal ranking models often tend to develop less data-intensive methods that rely on the judgment of professionals in the state agencies to integrate information into site rankings. A detailed state-by-state survey, beyond the scope of this study, would be needed to ascertain the relative utility of each of these somewhat different designs of ranking or rating sites.

8

COMPARING FEDERAL RANKING MODELS

THE DECISION-MAKING PROCESSES

The U.S. Department of Defense, U.S. Department of Energy, and U.S. Environmental Protection Agency have evolved three unique processes for choosing sites to remediate and for deciding the degree of hazard at each site. The committee reviewed these processes, specifically in an attempt to understand where each agency's model fits into their overall priority-setting process. Much less written information was available to the committee about the overall prioritization processes themselves than about the ranking models used in the processes. Written descriptions were often unavailable about the overall process and most of the information was obtained by questioning experts who visited the committee. With this caveat, the committee compares and contrasts in this chapter the procedures used by three federal agencies to assist in making decisions about remediation priorities.

Mandates for Remedial Action

A knowledge of the remedial action mandates of DOD, DOE, and EPA is critical to understanding their decision-making processes. EPA is required by federal law to be responsible for abandoned and active hazardous waste sites, underground storage tanks, a variety of other hazardous waste programs, and the integration of environmental laws, such as the Resource Conservation and Recovery Act (RCRA) of 1976 and the Toxic Substances Control Act (TSCA) of 1976, that impinge on hazardous wastes (e.g., polychlorinated biphenyls). EPA is also required by law to obtain funds for cleanup from responsible parties and from a tax on chemical products. Both of these fund-raising aspects are controversial. EPA is required to work with state environmental agencies and to serve in lieu of such agencies in some states. EPA is also expected to provide outreach for advice to local communities, partly in a manner mandated by the Superfund Amendments and Reauthorization Act (SARA) of 1986. In short, EPA is mandated to be involved in major decisions and is required to directly deal with a broad gamut of interested parties. Its decisions are open to public scrutiny at several stages in the process.

In contrast to EPA, DOE and DOD were initially protected from outside intervention by national security concerns. However, when contamination became an issue at DOD and DOE sites, these agencies were given responsibility by Congress for their own cleanup. Unlike EPA, their source of funds for work at sites is the federal government, not industry or state or local governments.

Site complexity is another important issue. DOD and especially DOE have some extraordinarily complicated sites that include hundreds, and sometimes thousands, of contaminated areas. While EPA is responsible for many more sites, these typically are easier to manage because they do not contain such large quantities and mixtures of radioactive and chemical wastes. Finally, closure of DOD bases and DOE sites imposes additional problems and complexities of scientific, social, economic, and political relevance on these two federal agencies.

Listing and Screening of Sites

DOD's, DOE's, and EPA's priority-setting processes have evolved to reflect the differences in their mandates, historical responsibilities, and site complexity. Superfund site nominations can come from a state, another federal agency, or other sources, which gives EPA minimal control over nominations. In contrast, DOE and DOD sites are primarily self-nominated by DOE and DOD personnel.

EPA was mandated by Congress to develop a method to pick at least 400 sites for the National Priorities List (NPL). Furthermore, Congress suggested factors to be used for choosing sites for NPL designation. The combination of a need to pick only a fraction of more than 50,000 potentially nominated sites and congressional input into the variables to be considered in the selection process led EPA directly to the development of a relatively simple mathematical screening model, the Hazard Ranking System (HRS), which is used to determine which contaminated sites will become part of the NPL and which will not. Since EPA must negotiate with responsible parties, states, and other stakeholders, it takes great care to document all of its data inputs on the assumption that its decisions might be legally challenged.

DOE and DOD have had much more control over which sites

to choose for analyses and remediation. The two agencies have divided their sites among special programs for rapid cleanup because of potential imminent hazards. For example, sites have been sorted into categories of higher priority cleanup before a base can be closed and less immediate action because they pose a less imminent but chronic threat. DOD (including the Army, Navy, and Air Force) and DOE exercise considerable control over the entire process. They can decide within their funding limitation whether or not to provide monitoring and cleanup funds to every potential site. Unlike EPA, initial scientific data inputs are not required to screen the large number of potential sites to obtain a smaller manageable set. DOE and DOD can collect information throughout the process at all sites. EPA in contrast does not have the resources to collect detailed data at tens of thousands of sites.

The DOD and DOE models described in this report reflect the greater control these agencies have over the decision-making process. DOE has the political independence and resources to develop a priority-setting method that requires extensive data gathering collection at every step in the process for its most hazardous sites. Because each site may contain mixtures of extremely toxic and radioactive as well as other types of wastes, DOE needs much information to make sound billion-dollar site cleanup decisions that encompass entire sites as well as modules within each site. Furthermore, the costs and benefits of remediating some DOE sites are potentially so substantial that DOE includes a cost and benefit section in its proposal formal decision-making model.

DOD's mandate leaves it somewhere between EPA and DOE with respect to the need for initial site-screening and extensive scientific data. DOD can provide money to base commanders for investigating every site, but not all DOD sites can be remediated fully at the same time. Consequently, DOD can collect data for prioritization at a later time in the overall process than EPA, but uses it within a simple ranking model to provide guidance to headquarters for decisions about the relative ranking of sites for cleanup.

Designation for Remediation

The combination of mandate and site complexity is also reflected in site-selection for remediation. Once a site is legally approved for inclusion on the NPL, more scientific data are gathered by EPA, usually in connection or in agreement with responsible parties and states. When these data have been gathered, they are used along with economic information to develop alternative remediation scenarios as part of the remedial investigation and feasibility study (RI/FS). These analyses, which lead to a Record of Decision for recommended remedial action, are sometimes challenged by stakeholders in court.

Responsible parties would like to use innovative technologies or methods that limit their costs and future liability. The local public is usually characterized as wanting sites remediated to their original uncontaminated condition, although Chapter 1 suggests this is not always the case. States frequently prefer remediating sites where the economic burden is on the responsible parties. Although the federal government contributes 90% of the total cost of remediation and the states contribute only 10% for abandoned sites to some states this is still burdensome during this period of little or no economic growth. Largely for this reason some states have had far fewer sites remediated than, for example, California and New Jersey, which have had more state funds available. States also tend to favor solutions that minimize continuing operating costs (e.g., pumping and treating contaminated groundwater) because states are responsible for 100% of the governmental share of the operating and maintenance costs. EPA must contend with all of the diverse interest groups when it makes decisions.

For many of the major sites, DOD and DOE have signed legal contracts with the states and EPA that mandate specific targets for cleanup of these selected sites. Many of the DOD and DOE sites are on the NPL and thus fall under EPA jurisdiction, but sites with such legal agreements clearly take priority over other sites.

DOD and DOE have made some formal attempts to include local officials and concerned citizens in their decision-making process, but their mandate to consult with other interested parties is not nearly as strong as it is for EPA.

Comparative Scoring Exercise Of Federal Ranking Models

Objective and Background

Mathematical models are being used or are under development by EPA, DOE, and DOD to serve as aids in the overall priority-setting process described above for determining which sites to cleanup first. The site rankings resulting from model calculations are often thought to be a major determining factor in selection of sites for cleanup, but this is not always the case. Many other considerations enter into these decisions, as already discussed. Nevertheless, mathematical models can help to evaluate complex factors thought to be important, including risk to public health and the environment. The committee spent most of its efforts in a review of the models being used or developed by the three federal agencies; those models tended to be much better documented and more subject to scientific scrutiny than the overall priority-setting processes themselves.

To learn about the relative performance of models used for ranking sites, the committee felt that it was essential to apply DOD's DPM, DOE's MEPAS, and EPA's HRS to a common set of sites. The committee performed its analysis with the awareness that these models have different purposes, features, and data requirements. The scoring exercise helped to familiarize the committee with the models' input data requirements and operating constraints, and characteristics that contribute to similarities and differences in model outputs. For the five hazardous waste sites selected, the exercise also helped the committee to determine whether the three models produced similar relative hazard rankings based on potential exposures to humans.

The purpose of this scoring exercise was not to determine which model is right or wrong nor whether one ranking method is better than another, but to compare outputs obtained using input data developed from the same set of contaminated sites. Table 8-1 summarizes the features of each model. Table 8-2 identifies the environmental transport pathways addressed by each model. Note that although features of DOE's Environmental Restoration Priority System (ERPS) are shown in Table 8-1, it was not included in the scoring exercise because, as discussed in Chapter 6, its design and application are very much different from DPM, MEPAS, and HRS.

Differences in the models' features make an exact comparison of the models very difficult. For example, the HRS and the DPM are scoring systems, whereas MEPAS is a fate and transport (FaT) model-based system. Scoring systems assign arbitrary numerical values to FaT and other parameters that characterize the site. These numerical values are then combined by an arbitrary algorithm and normalized to yield a site score. The process of contaminant migration from a site is not directly determined in a scoring system. Conversely, FaT model-based systems use contaminant mass balances in complex mathematical formulas to predict and quantify the migration of contaminants to potential receptors. MEPAS includes FaT algorithms for site contaminants and an arbitrary final numerical score, *hazard potential index*, to scale-quantify the risk to human health. Although different in nature, the three models might be expected to produce similar relative rankings when applied to the same set of hazardous waste sites—that is, to give the same indication of which sites produce the higher health risk.

Each agency was provided with a common set of site descrip-

231

TABLE 8-1 Comparison of Features of EPA's Hazard Ranking System (HRS), DOD's Defense Priority Model (DPM), DOE's Multimedia Environmental Pollutant Assessment System (MEPAS), and DOE's Environmental Restoration Priority System (ERPS)

Feature	HRS	DPM	MEPAS	ERPS
Purpose				
Screening	Yes	Yes	Yes	Yes (for emergencies)
Alternative evaluation	No	Possible	Yes	Yes
Resource allocation	No	No	No	Yes
Types of sites covered	All	DOD	DOE	All
Contaminant type				
Radioactive	Yes	Yes	Yes	Yes
Nonradioactive	Yes	Yes	Yes	Yes
Risks				
Human health	Yes	Yes	Yes	Yes
Environmental	Yes	Yes	No	Yes
Socioeconomic	No	No	No	Yes

Cost estimates of remediation					
Alternatives	No	No	No	No	Yes
Dynamic tracking	Suggested, but not appropriate	Could be adapted to do so	Could be adapted to do so	Could be adapted to do so	Yes (year to year)
Required QA/QC	Yes	Yes	Yes	Yes	Yes
Explicit value-preference weights	No	Yes	No	No	Yes
Transparency	Fair[a]	Yes	Yes	No	No
User-friendly	Fair[a]	Yes	Yes	Yes	Yes
Security features	Yes	No	No	Yes	No
Process of model review					
Peer review	Yes	Yes[b]	Yes	Yes	Yes
Public participation	Yes	Some	Yes	Yes	Yes
Comments	Yes	Yes	Yes	Yes	Yes
Internally consistent logic and math	Yes	Yes	Yes	Yes	Yes

TABLE 8-1 (continued)

Feature	HRS	DPM	MEPAS	ERPS
Documentation	Adequate	Inadequate	Adequate	Inadequate
Validation attempted	Yes	No	Yes	Yes

[a] Revised HRS is less transparent and less user-friendly than original HRS.
[b] Performed by the committee.

TABLE 8-2 Environmental Transport Pathways Addressed in EPA's Hazard Ranking System (HRS), DOD's Defense Priority Model (DPM), and DOE's Multimedia Environmental Pollutant Assessment System (MEPAS)

Pathway	HRS	DPM	MEPAS
Air-vegetation	No	No	Yes
Soil-vegetation	No	--[a]	Yes
Water-aquatic biota	Yes	--	Yes
Terrestrial animals	No	--	Yes
Air-water	No	Yes	Yes
Water-air	No	Yes	Yes
Air-soil	Yes	Yes	Yes
Water-sediment	No	No	Yes
Sediment-water	No	No	Yes
Soil-water	Yes	Yes	Yes
Soil-groundwater	Yes	Yes	Yes
Groundwater-soil	No	No	Yes
Groundwater-surface water	Yes	--	Yes

[a]Not determined.

tions, narrative, background, and data, and they were asked to run their respective models and to provide the committee with the scoring results. The sites were selected in consultation with representatives of EPA, DOD, and DOE from a database describing actual sites. In selecting the sites, an effort was made to obtain a broad representation of site characteristics such as contamination and site type, potential for human exposure, and potential for ecological impact. To ensure comparable applications, extensive interagency communications occurred. Upon completing the model runs, each agency summarized its results and submitted reports to the committee. The numbers 1-5 were used instead of site names.

Ranking Hazardous Waste Sites

Description of the Five Hazardous-Waste Sites Included in the Scoring Exercise

Site 1

Site 1 is a sanitary landfill located on glacial till that was used for solid waste disposal. No engineered liner exists below the landfill. Before 1981, disposal of waste at the site appears to have been uncontrolled and largely undocumented. A vertical french drain located on one corner of the site was used from 1969 to 1978 to dispose of approximately 29,000 gallons of liquid hazardous waste, including various organics, waste oil, diesel fuel, kerosene, and fluids containing polychlorinated biphenyls (PCBs). The french drain was clay capped in 1979. No waste was removed prior to capping. Smaller french drains at other locations in the landfill might have been used for disposal of asbestos from demolition and reconstruction projects. The amount of asbestos disposed in the landfill fluctuated from 615 cubic feet in 1984 to 2,000 cubic feet in 1987. The landfill is located 0.3 miles from the nearest resident community and industrial facilities, 0.01 miles from wetlands located in a nature preserve, and 0.25 miles from agricultural lands. Demographics show 3,300 people reside within a 1-mile radius of the site, 12,000 people within a 2-mile radius, and 39,600 people within a 3-mile radius. Approximately 8,000,000 people surround the site within a 50-mile radius.

Site 2

Site 2 is a settling basin that was designed and operated to allow wastewater to percolate through the soil at the sides and bottom. Soil beneath the settling basin has been classified as sand. During its operation, it received approximately 3,000 cubic meters per day of wastewater. Effluent from the site was discharged into a tributary of a nearby creek. Liquids (containing nitrates and organic solvents) often overflowed, entered the groundwater, and cropped out into the nearby creek and a downstream river and lake. The primary source of effluent sent to the basin was characterized as electroplating waste. The site is surrounded by forested countryside and nonforested areas comprising lowland hardwood swamps, sand hills, old agricultural fields, and aquatic areas. Two large population centers are within 25 miles of the site; one has a population of 25,000 people. The population density in the counties surrounding the site ranges from 23 to 560 people per square mile. Approximately 583,000 people surround the site within a 50-mile radius.

Site 3

Site 3 is a drum storage yard site with 78,000 drums stored on an asphalt pad. This site is located by a stream that discharges into a river. The subsurface below the site is characterized as a layer of clay with varying content of silt, sand, and rock fragments that overlies limestone and dolomite rock. The drums contain inorganic and organic waste sludge that was collected from a retention basin and holding pond. The outside of the drums show signs of internal corrosion. Sludge contained in 45,000 of the drums has been stabilized in concrete grout; those drums are not thought to contribute to fugitive emissions. Another 32,000 drums contain raw sludge. Liquid sludge has leaked from some of the drums onto the asphalt pad and possibly into the surrounding soils. Solid sludge might also be escaping from some drums. 8,000 drums have been drained of free liquid and moved elsewhere on-site. The site is surrounded by predominantly rural land consisting of forested, agricultural and industrial areas, with two major population centers of approximately 28,000 and 180,000 people within 25 miles. Approximately 820,000 people surround the site within a 50-mile radius.

Site 4

Site 4 is a hilltop building complex used for material testing and high-explosive diagnostic work. The site is surrounded by hills, ridges, ditches, and gullies. The subsurface under the site is characterized by considerable horizontal and vertical variability. Volatile organic compounds (VOCs), primarily trichloroethylene, have been pumped through above-ground pipes within the complex. Leaks in the pumping-station building, the valve system outside the pumping-station building, and the work areas have led to VOC releases into the environment. Spills within the complex have been washed into floor sumps that drained into the septic system. The septic system leach field is also a source of VOC release. The population near the site is fewer than one person per square mile. The site is surrounded by grassland. Land use around the site is primarily agricultural and recreational. Approximately 4,140,000 people surround the site within a 50-mile radius.

Site 5

Site 5 comprises waste pits which received various types of metal-containing slurry or dry solid waste. Some pits were lined with compacted native clay and others had rubberized elastomeric membranes. Next to the waste pits (no longer in use) is a burn pit that was used for disposal of chemicals and other combustible waste. In the same area, a clearwell was operated as a settling basin for process waste and storm water runoff. Leakage through the pit bottoms and runoff into a nearby creek has led to groundwater and surface water contamination. The release of these wastes into the environment has occurred for over 35 years and has led to contamination of air, soil, surface water, and groundwater. The site is surrounded by forested countryside. Within a 5-mile radius of the site, an estimated population of more than 24,000 people resides. Approximately 2,500,000 people surround the site within a 50-mile radius.

Model Results and Comparison of Scores

The following is a detailed analysis of the scores and an evaluation of the comparative scoring exercise. Table 8-3 lists the scoring results and descriptive statistics for the three models. The scores are displayed graphically in Figure 8-1 as an aid to visual comparison of the outputs.

TABLE 8-3 Results from Scoring Five Sites Using DOD's Defense Priority Model (DPM), EPA's Hazard Ranking System (HRS), and DOE's Multimedia Environmental Pollutant Assessment System (MEPAS)

	Model Score by Site Number					Descriptive Statistics		
	1	2	3	4	5	M[a]	SD	Ratio
DPM	45.4	28.6	19.9	15.2	33.5	28.5	11.8	0.4
HRS	73	28	1.2	2.2	4.6	21.8	30.7	1.4
MEPAS	53	58	33	23	56	45	16	0.4

[a]M = Mean; SD = Standard deviation; Ratio = Standard deviation/mean.

Source: L. Zaragoza, EPA; data provided to the committee September 4, 1992.

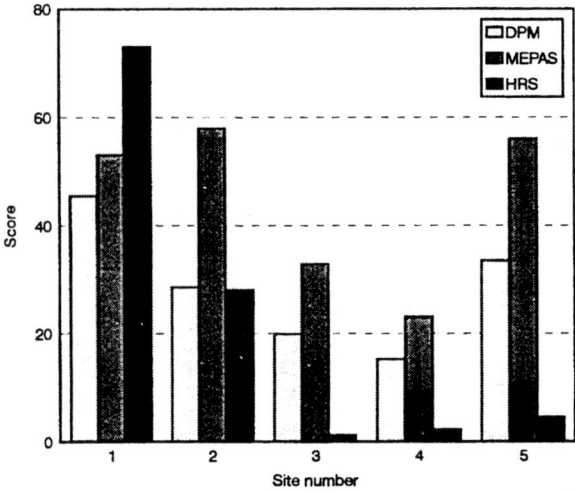

FIGURE 8-1 Graphical presentation of scores

Upon preliminary inspection of Figure 8-1, it appears that the three models give similar hazard ranking predictions. However, upon closer inspection and interpretation, a different result emerges. Although it is argued that the models were developed for different purposes and use different data inputs, among other things a stated or implied purpose of each model is to rank relative potential hazards. The three models do not agree as to which site of the five poses the highest hazard. ("Hazard" is a generic term for all three models; "hazard level" corresponds to the magnitude of the site scores that are used for screening, priority setting for cleanup, or inclusion on the NPL.) Table 8-4 presents the distribution of site rankings obtained from each model arranged according to each site's hazard level.

DPM and HRS rank Site 1 highest, while MEPAS ranked it much lower as moderate-low. There is no uniform agreement as to the next-to-highest hazardous site. DPM and MEPAS rank Sites 2 and 5 as moderate, and HRS ranks Site 2 as low and Site 5 as very low. All three models rank Site 3 as low or very low and Site 4 is ranked as very low. It is disconcerting that the models agree better on those sites that might have a low hazard level than those that might have a high hazard level.

TABLE 8-4 Distribution of Rankings of Five Sites According to Output from DOD's Defense Priority Model (DPM), EPA's Hazard Ranking System (HRS), and DOE's Multimedia Environmental Pollutant Assessment System (MEPAS)[a]

Site's Hazard Level	DPM	HRS	MEPAS
Very high	--[b]	--	--
High	1	1	--
Moderate-high	--	--	--
Moderate-medium	2, 5	--	1, 2, 5
Moderate-low	--	--	--
Low	3	2	3
Very low	4	3, 4, 5	4

[a]Numbers refer to site identification.
[b]Denotes no site ranked with this hazard level.

Is it significant that for Site 5 the MEPAS and DPM rankings are moderate-medium while for HRS the same site is ranked as very low? Based on the data presented in the site description, this site would seem to very objectionable. That the HRS model would erroneously rank this site as a very low hazard—if this is indeed an error—suggests that it, and likewise the other models might have misranked possibly numerous other sites.

As noted previously, although the models were developed for slightly different purposes, identifying high-hazard sites was an underlying feature in all three models. All therefore should have this basic capability, the committee believes it is not appropriate here to apply statistical arguments or other such mathematical formalisms to compare the model predictions in an absolute sense. With respect to ferreting out high-hazard sites, all three models should be absolutely in agreement rather than statistically in agreement some percent of the time.

The developers of the respective models scaled the scores arbitrarily such that a score of 100 represents the very highest hazard sites and 0 represents the clean sites. The highest numerical score was 73, obtained by HRS for Site 1. The lack of high scores does not necessarily mean that all five sites selected for scoring happened to be relatively low in hazard. For example, an HRS score of 28.5 or higher places a site on the NPL.

Normalized scores for a model were obtained by dividing the individual site score values by the model average for all sites. A normalized graphical comparison of scores appears in Figure 8-2. On this basis, the numerical congruence of DPM and MEPAS is very good, while HRS follows the general trend of the others. Since the general site ranking protocol embedded in each model is basically the same, as noted in the separate chapters on each model, this congruent behavior is to be expected. Although there is congruence of the DPM and MEPAS scores, MEPAS uses detailed transport and fate algorithms while DPM and HRS use a simpler structured-value approach to model transport and fate elements in arriving at scores. Considering the congruence of DPM and MEPAS, it appears that the simpler approach might be sufficient when relative scores are desired. More comparative testing should be done to explore this hypothesis.

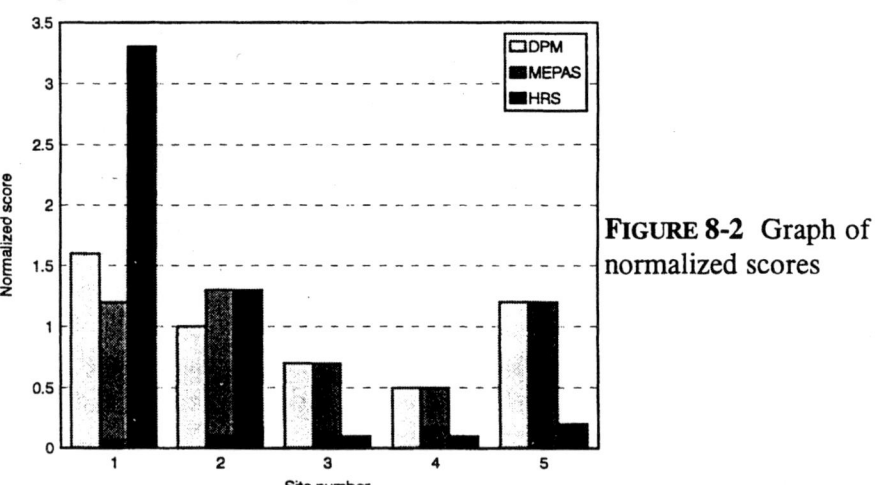

FIGURE 8-2 Graph of normalized scores

DPM and MEPAS have different methods by which they identify contaminants associated with a site. In general, applications of MEPAS require that the contaminant data most directly associated with the waste site be used in preference to data indirectly attributed to the site. For most MEPAS applications, the data judged most readily attributable to a single waste unit are estimates of the quantities of contaminants placed at the waste site (i.e., the contaminant inventory). Contaminant concentrations in environmental media near the waste site (e.g., groundwater contaminant concentrations) can only be indirectly associated with a specific waste site because multiple potential contaminant sources usually exist in the vicinity of the site. Thus, although MEPAS has the capability to use groundwater and surface water concentrations as source terms, users often choose to use inventories as source terms. Conversely, DPM relies heavily on *observed* concentrations of contaminants in the air, surface water, groundwater, or soil near the waste site to be scored. If no sampling has been done at the site, DPM assumes a worst-case approach by assuming all chemicals in the inventory have contaminated the groundwater and surface water pathways, but not for the pathways of atmospheric transport. Thus, DPM site scores are based on contaminants identified by sampling the surrounding environmental media, whereas MEPAS scores can be and often are based on contaminants identified in contaminant historical inventories. A discrepancy arises when sampling data used for DPM scoring indicate the presence of contaminants that are not identified in the contaminant inventory used for MEPAS scoring. The presence of additional contaminants not identified in the inventory might result from the identification of contaminants that have migrated from other sources or from incomplete contaminant inventories.

In several instances, MEPAS and DPM identify the same contaminants of primary concern, yet produce conflicting rankings of secondary contaminants. The committee suspects that this discrepancy results from the use of differing benchmark values as known standards with which to compare the estimated intake.

MEPAS bases its index of relative risk, the Hazard Potential Index (HPI), on toxicological indicators of potential harm—the slope factor for carcinogens and the reference dose for noncarcinogens—thereby yielding a value of cancer risk and a hazard quotient for carcinogens and noncarcinogens, respectively. Conversely, DPM derives its benchmarks from regulatory limits (e.g., maximum contaminant levels) and does not differentiate between carcinogens and noncarcinogens. Some of the DPM's benchmarks are derived from the same toxicity values MEPAS uses, since regulated values are not available for every chemical for every pathway. Although MEPAS and DPM are expected to correctly identify a specific contaminant as the primary risk producer at a site, the discrepancy created by use of regulated values could contribute to the different relative rankings of secondary contaminants.

Unlike DPM and HRS, which account for potential harm to both human health and the surrounding ecosystems, MEPAS considers only risks to human health.

Site 1

The DPM and MEPAS evaluations of Site 1 differ in that MEPAS identifies the atmospheric transport pathway as the pathway producing the highest risk, whereas DPM identifies the groundwater pathway as the pathway contributing most to the site score, followed closely by the atmospheric pathway. The high groundwater score of DPM can be attributed to the fact that DPM identified and scored the site using groundwater contaminants that were not identified in the MEPAS source term. With consistent contaminants, both models probably would have identified the same pathway. The evaluation of the atmospheric pathways uses a consistent set of contaminants, and consequently, the relative rankings of contaminants evaluated in the atmospheric pathway were in good agreement.

For this site, MEPAS evaluated risk from surface water contamination resulting from the recharge of contaminated groundwater but did not evaluate the pathway of overland runoff to surface water. DPM evaluated surface water hazard from overland runoff, but not from contaminated groundwater recharge. DPM considers groundwater recharge of surface water in the groundwater pathway rather than in the surface water pathway. Neither model run identified the surface water pathway as a significant contributor to the overall site risk.

While the DPM and the MEPAS evaluations scored the surface water pathway low, the HRS scored the surface water pathway with the maximum points. This high score was due to an observed contaminant release to a watershed that has a hatchery within 15 miles. An observed release in this situation automatically receives an HRS moderate target score of 120 points for a food chain threat and it increases the surface water pathway score since at least one highly toxic and bioaccumulative substance was present at the site.

Site 2

For Site 2, DPM and MEPAS identified the groundwater pathway as the pathway of primary concern; HRS scored groundwater as low. Like Site 1, several groundwater contaminants were identified and used in the DPM scoring that were not evaluated with MEPAS, because prior modeling of this site with MEPAS relied on the contaminant inventory as the source term.

The primary reason that the HRS scored the groundwater pathway low is because of the relative emphasis placed on linking contaminants and targets. HRS evaluated multiple aquifers at Site 2. No releases were observed in the highest scoring aquifers, which had a fairly high target score, and consequently, the pathway score was substantially lowered. The aquifers with observed releases had few targets and received very low scores.

MEPAS evaluated a groundwater-to-surface water pathway for this site while DPM evaluated an overland runoff to surface water pathway. Again, DPM scored this site using sampling data that identified potentially hazardous contaminants that were not identified by the users of MEPAS as part of the contaminant inventory. This discrepancy explains why the surface water pathway was more significant in the DPM scoring than in the MEPAS scoring.

Both MEPAS and DPM evaluated atmospheric transport of contaminants from this site, but the relative rankings of the contaminants varied. This difference is probably due to the differing toxicity values used by each of the models to produce an index of relative risk.

Site 3

For Site 3, both DPM and MEPAS identified the overland runoff to surface water pathway as the pathway of primary concern. The groundwater pathway was found by both models to be less significant, and both models found little or no risk from atmospheric transport pathways. Due to these similarities, both models ranked this site fourth in relation to the others, indicating relatively consistent predictive abilities of the two models.

HRS evaluated the surface water pathway as low. An explanation for this difference is that the DPM accounts for the presence or absence of a sensitive environment and the distance to the sensitive environment while HRS does not. HRS weights the sensitive environment based on surface water characteristics, such as stream flow, and not on distance (unless there is documented actual contamination of the sensitive environment, in which case the HRS score is not flow-weighted).

Site 4

MEPAS and DPM ranked Site 4 lower than all of the others. Despite this similarity, the models differ in the proportion of risk attributed to each contaminant transport pathway. The DPM score is driven by the groundwater pathway, whereas MEPAS identifies low but roughly equal risks from both the groundwater and atmospheric pathways. Because the atmospheric components of other site scores are in good agreement and both models use the same soil concentration of trichloroethylene (1 µg/g) to estimate risk from the atmospheric pathway, the reason for this discrepancy is unclear.

Because the only nearby stream is intermittent in nature and is recharged by groundwater upgradient from this site, MEPAS did not consider the surface water pathway a viable pathway for contaminant migration from Site 4. No surface water sampling data existed for the site. DPM assumed a worst-case approach by assuming all surface soil constituents could contaminate the surface water. The fact that the surface water pathway in DPM only contributed 1% of the overall site score even when based on the worst case confirmed the MEPAS assumption that risk from the surface water pathway was negligible.

HRS does not consider intermittent streams to be surface water bodies unless the area has less than 20 inches of mean annual precipitation. Because the nearest permanent water body was greater than 2 miles away, HRS did not score the surface water pathway for this site.

Site 5

Although MEPAS and DPM ranked Site 5 second out of the five sites scored, the primary risk-producing pathway was different for each model. MEPAS identified the surface water pathway as the pathway contributing the highest risk, whereas the DPM identified the groundwater pathway as the primary contributor to the overall site score. HRS scored the groundwater pathway low since sampling data did not indicate that contaminants had actually reached the target and no large drinking water supply well was near the site.

MEPAS and DPM identify PCBs (Arochlor 1254) and 1,1-dichloroethane as the contaminants of primary concern in the surface water and groundwater, respectively. The relative significance of secondary contaminants differ in all pathways. As described above, the committee attributes the difference in relative significance of secondary contaminants to differing methods of deriving benchmarks.

Scoring Exercise Conclusions

The scoring of these five sites indicates the following

- Scores obtained from all three models generally follow the same trend from site to site; however, in each model this trend results from very different reasons.
- Model hazard rankings do not agree on which site is the highest or next to the highest.
- Model hazard rankings do agree on which sites are a relatively low hazard.
- A fundamental difference in the recommended use of con-

taminant data results in a different set of contaminants being evaluated by each model, which leads to a discrepancy in the differences in site scores.

• A fundamental difference in the recommended use of data to weight different environmental transport pathways also leads to a discrepancy in the differences in site scores.

• Even when relatively consistent site scores are produced, the dominant risk-producing contaminants and transport pathways usually differ between models.

• Differences in evaluation of site data, rather than model structure, appear to be the major factor leading to variance in site scores. However, DPM and HRS evaluate potential risk to human health and the environment, whereas MEPAS evaluate potential risk to human health only. The effect of such a difference on the scoring results is unknown.

General Conclusions

Comparing the performance of ranking models is a useful exercise. It should be used on a regular basis in the future to compare the performance of newly developed or modified models against the output of established ones. The results of more complex and expensive models should be compared with older and simpler models. A set of reference sites should be established for use in developing input data for a wide range of hazardous potential (high to low). Using a set of about 12 dissimilar sites would help ensure that the models are compared on the basis of a broad range of site characteristics. The approach would be useful for checking whether revised algorithms are performing as expected or whether the models can discriminate, in a numerical sense, among various degrees of potential site hazards in the range of high to low. Objective criteria for these comparisons, however, are needed both in a relative sense (comparing one site with another) and an absolute sense (identification of most threatening sites).

9

TOWARD A UNIFIED NATIONAL APPROACH

The need for a unified national approach for setting remedial action priorities for sites contaminated by hazardous substances became evident during the committee's review of the different approaches currently used. Many steps in the present processes for setting priorities are not open to public scrutiny, and the ranking models used in those processes that were reviewed by the committee often lacked sufficient scientific rigor and validation. The situation calls for the development of a scientifically and environmentally sound, publicly acceptable, and consistent process that is commensurate with the enormous costs required for adequate site remediation (let alone complete restoration). This chapter discusses the advantages and disadvantages of a unified national process for setting priorities and proposes one such unified process.

Advantages and Disadvantages of a Unified Approach

During the past 15 years, a set of complex institutional arrangements have evolved in the United States for selecting and managing hazardous waste-site remediations. EPA, DOD, DOE, the states, and legally responsible parties all play major roles. Other federal agencies, local governments, environmental and legal consultants, the mass media, professional and industrial organizations, the judicial courts, and environmental advocacy groups are also important participants.

At the present time, there is no consistent relationship between the hazard present at a site and the process by which it is screened and evaluated for remediation. For example, EPA works closely with DOE and the states to develop plans to remediate DOE sites. Other sites are the responsibility of a single agency. Clearly, the processes involved in evaluating and remediating sites far outstrip the relatively simple model that the creators of Superfund had envisioned.

The remediation process is also much more expensive than expected. The number of high-priority sites is already three times more than the 400 required by the original Comprehensive Environmental Resource, Compensation, and Liability Act (CERCLA) of 1980, and current cost estimates in the hundreds of billions of dollars for remediating all active and inactive contaminated sites during the next 30 years dwarfs the original and amended CERCLA authorization of about $10 billion. The committee's analysis of the EPA, DOD, and DOE ranking models shows major differences in the history of site ranking model development, differences in the underlying logic and science supporting model development and testing, and differences in the use of science-based results to allocate scarce resources. In short, each agency has developed its own unique protocol. However, the use of independent unique processes might not be in the best interests of the United States as a whole, especially during a period of pressure to accomplish more with fewer resources.

The committee recommends that the United States considers a common process of scientific analysis of sites to replace the existing multiplicity of approaches. Three alternative strategies for reducing inconsistencies have been identified, based respectively on: greater consultation, scientific consistency, and decision-making consistency. Greater consultation is the least intrusive to existing agency approaches. EPA, DOD, and DOE would form an interagency task force to review existing mathematical site-ranking methods and determine how the agencies can better share data, expertise, quality control, validation procedures, and other scientific and mathematical information. For example, data availability and data quality are major concerns. Any agreement about the kind of data that should be gathered and how its quality should be addressed can only be helpful to all agencies in the long run. The three agencies would also share their informal processes for including local social and economic concerns and their processes for communicating with tribal, state and local governments, interested parties, and the general public. Greater consultation would change existing site-ranking, decision-making, and budgetary processes only if and insofar as the agencies agreed to make changes and the federal administration agreed to the change.

The second alternative, a strategy of scientific consistency, requires that each site be subjected to the same scientific protocol for evaluating health and safety, environmental impact, and economic costs and benefits. For example, whatever scientific protocol is applied to a DOD site in Wyoming would also be applied to a DOE site in Maryland and to an EPA site in California. The committee believes that such use of a unified scientific approach would make the scientific input into the political process more explicit and more thorough. This uniform scientific process would be embedded as a common component in each agency's larger process of deciding on resource allocations for site identification, ranking, and remediation. A unified national approach that standardizes the scientific elements of remediation decision-making would not replace or diminish the political parts of the process—that is, the parts that require bargaining among the major parties.

Decision-making consistency defines the third and most unifying alternative, incorporating scientific consistency, it goes further to add process and geographical consistency. All agencies would apply the same scientific protocol to each of its sites and would allocate remediation resources on the basis of the outcome of that protocol. In other words, resources would not be influenced by which agency was responsible for the site. For example, the remediation of a solvent spill on factory grounds in Illinois would be treated in the same way as solvent spills at a military base in Arizona or at a DOE facility in Ohio. Priority would be assigned by a central interagency group, not exclusively by the parties currently charged with remediating the site. The committee recognizes this alternative would involve major reorganization of responsibility, authority, and budgetary resources among the three major federal agencies charged with cleanup of hazardous waste sites.

Although all three approaches merit consideration, this chapter focuses primarily on scientific consistency because the committee's charge and expertise concentrate on the scientific factors that influence hazardous waste site assessment and management. First, the advantages and disadvantages of a unified national scientific and decision-making approach will be discussed from five policy perspectives, and then a particular proposal for a unified national priority-setting process will be presented.

The five policy perspectives addressed in the following discussion of advantages and disadvantages of a uniform scientific approach to aid in decision making for site remediation are: protection of health and the environment, investment of funds, organization acceptability, consistency, and adaptability.

Toward a United National Approach

Protection of Health and Environment

The major advantage of requiring scientific consistency is that it would focus previously fragmented efforts, leading the best scientists and policy analysts to a concerted collaboration in developing the models necessary to give the most credible estimates of health, environmental, and welfare (e.g., land value) impacts and costs. This should, in turn, lead to the best scientific protocol now possible as a basis for decision-making, and to the best possible programs of data gathering and research for continued improvement of the protocol. Scientists from EPA, DOD, DOE, and the Agency for Toxic Substances and Disease Registry (ATSDR) would all be involved in selecting the best scientific approach for estimating risk to human health, human welfare, and the environment. For example, instead of each agency heading in its own direction to assess the impacts of contaminants on water quality and on the people who drink the contaminated water, the agencies could combine their efforts as well as get input from scientific experts outside the agencies and from public interest groups. They could start by scrutinizing their existing approaches in order to isolate those discrepancies in the assumptions, values, and scientific methods used in the models that result in different ratings and rankings for the same set of sites (see Chapter 8). They could establish a uniform definition of "site," join together to build a formal site discovery program, and formulate uniform criteria for an emergency cleanup. A better decision-making protocol would also help ensure that protection for human health and the environment is adequate.

A disadvantage is that the existing methods embody the investment of a great deal of time and money. Although these methods would undergo gradual refinement, the development of a national protocol might lead to a more rapid displacement for some of them. Resources would be needed to retrain technical personnel to implement methods of the protocol.

Investment of Funds

Only one of the methods considered by the committee, DOE's priority-setting approach, explicitly incorporates a consideration of costs and benefits. In light of the enormous costs of alternative site-remediation processes, a single consistent national process that formally includes representations of economic costs and benefits would be advantageous to decision-makers. In essence, such a process would recognize the reality that costs and benefits are always factored in some way into decisions. The committee believes that an explicit treatment of costs and benefits of remediation would increase the credibility of the process by providing estimates that could be compared with actual site costs and benefits—in a sense, a kind of cost and benefits accounting. A clearly documented costs and benefits protocol element should greatly reduce the possibilities for inadvertent or intended skewing of cost and benefit considerations for reasons that have nothing to do with hazard or remediation outcomes.

One disadvantage is the possibility of principled opposition by those who believe that economic costs and benefits should not be a consideration in protecting public health and environment, and would be distressed by such explicit consideration of costs and benefits. Also, it would take time to build an economic protocol that would gain broad acceptance to practitioners, interested parties, and theoreticians.

Organization Acceptability

There are obvious organizational disadvantages for the major federal agencies, and perhaps their established consultants, all of whom would have some incentive to resist the imposition of a national scientific approach. This resistance could lead to a long and drawn out effort to develop a unified system. The existing priority-setting processes have led to legal agreements that already promise remediation, so that signatories would feel threatened by the imposition of an approach that may be perceived as trying to set the clock back.

The previous point is countered by assessments showing that limitations on financial resources will probably make it infeasible to fulfill all existing legal agreements concerning site remediation. If and when this situation becomes recognized reality, a single consistent process will be invaluable in aiding the major federal agencies to bargain with each other, as well as with the states, local governments, private sectors, citizens groups, and other stakeholders because all will be negotiating from the same data base and on a more level playing field. Ideally, such a uniform approach would lead the federal agencies to develop a joint strategic plan for remediation under a variety of resource-constrained scenarios. In other words, it would lead to a hierarchical set of objectives to accommodate financial and technical limitations and advances.

Consistency

There appears to be geographical inconsistency in the distribution of site-remediation resources. Some states have strong environmental protection programs and others do not. The result is a situation where not every American community and ecosystem is protected on a consistent basis by the current priority-setting or resource-allocation methods. A major advantage of the uniform scientific approach is that every analysis will treat every person the same and every forest the same, regardless of whether they are located in an urban area of New Jersey or Louisiana, or a rural area of Maine or Arizona. Going further, decision-making consistency would allocate remediation resources on the basis of costs, benefits, and need for cleanup rather than on the basis of the ability of a responsible party, state, business, or federal agency to pay.

A disadvantage stems from the fact that many state governments under the policy of new federalism have been given back much of the authority to choose how they spend their limited resources. Many can be expected to oppose a national process that would ascribe high benefits compared with costs for remediation that is not a high priority for them. Likewise, under decision-making consistency, federal agencies can be expected to resist a decrease in their authority to decide priorities for their sites.

Some state opposition might be overcome by allowing states to use their own method of assessing cleanup priorities as long as they also take into account the results produced by the national method as a baseline. In other words, states would, at a minimum, use the national method. They could also use their own method and therefore would have two sets of information upon which to inform their discussion. In addition, states could be given the right to change the weights assigned to different environmental elements of models used in a national priority-setting approach (e.g., groundwater, forests, and housing). Each state could change the outcome for its sites by changing the weights of importance, but these changes would have to be explicit, documented as to justification and process of derivation, and open to public scrutiny. A similar accommodation might be made for federal agencies. The state and federal agencies would receive the same information and be able to use it to guide, but not dictate, their final decisions.

Adaptability

The committee believes that a national approach to setting priorities would better accommodate changes in the scientific, technological, economic, and political processes in the United States and abroad than do the existing multi-organizational processes. However, some states are more innovative and adaptable than is the federal government. These few states would have the disadvantage of waiting for a federal process that would probably be slower. On the other hand, these states would doubtless take the lead and press the federal government and other states for change.

Much has been learned about evaluating risks from site contamination over the past 10 years, and new technologies have been developed that make site remediation more economical. In addition, the limits of the nation's technical and economic ability to remediate sites have become more widely recognized. Because of the enormous cost of remediation to attain unrestricted land use in many instances, and the technical near-impossibility of attaining this goal in others, there is growing opposition to the original goal of complete restoration of sites (see Chapter 1).

Furthermore, even though remediation is likely to go on for decades, many resources will be allocated throughout the 1990s as well as beyond. If the United States is ever to adopt a uniform national scientific and decision-making process, it makes sense to do it soon.

PROPOSED UNIFIED NATIONAL PROCESS FOR SETTING PRIORITIES

The three major federal agencies involved in site restoration use different approaches for evaluating site risks and setting site cleanup priorities. It is extremely difficult to compare the consistency in degree of cleanup and level of protection being provided by the different agencies. Nor have the risk-based procedures currently in use been adequately validated, partly because the effort, expense, and time to do so appears excessive. However, the committee considers such validation attempts essential for a program presenting such enormous costs, and thus demanding maximal assurance that funds are being spent wisely. With use of a unified scientific approach by all agencies, the cost and effort in development and validation can be shared, and a national consensus on the directions to be taken becomes more supportable and likely.

The committee proposes serious consideration of a three-tiered unified procedure for setting priorities for hazardous waste site remediation. The approach proposed draws heavily on procedures already being used either explicitly or implicitly by state and federal agencies, including EPA, DOD, and DOE, so that no radical change in thinking or development is required. A general outline of this three-tiered approach is presented in Figure 9-1.

The first tier of the unified approach embodies a procedure for screening candidate hazardous waste sites. Here, a site is evaluated simply to determine whether it (1) should be moved to the second tier for more detailed characterization, (2) should be eliminated from further consideration, or (3) should be held for further analysis or characterization so a decision can be made in a timely manner to move it to Tier Two or eliminate it from further consideration. The decisions in Tier One would necessarily be based on limited data regarding the degree of hazard to human health and the environment. The Hazard Ranking System of EPA and DOD's Defense Priority Model are examples of models that have been developed for the type of evaluations that would be performed in Tier One.

In Tier Two, health and environmental risks and remediation costs are estimated. A detailed site investigation would be conducted to determine the extent of contamination present at a site and the various environmental media and populations that might be affected by the contaminants. The data obtained here should be adequate to conduct a formalized assessment of the relative risk posed to human health and the environment. The objective here

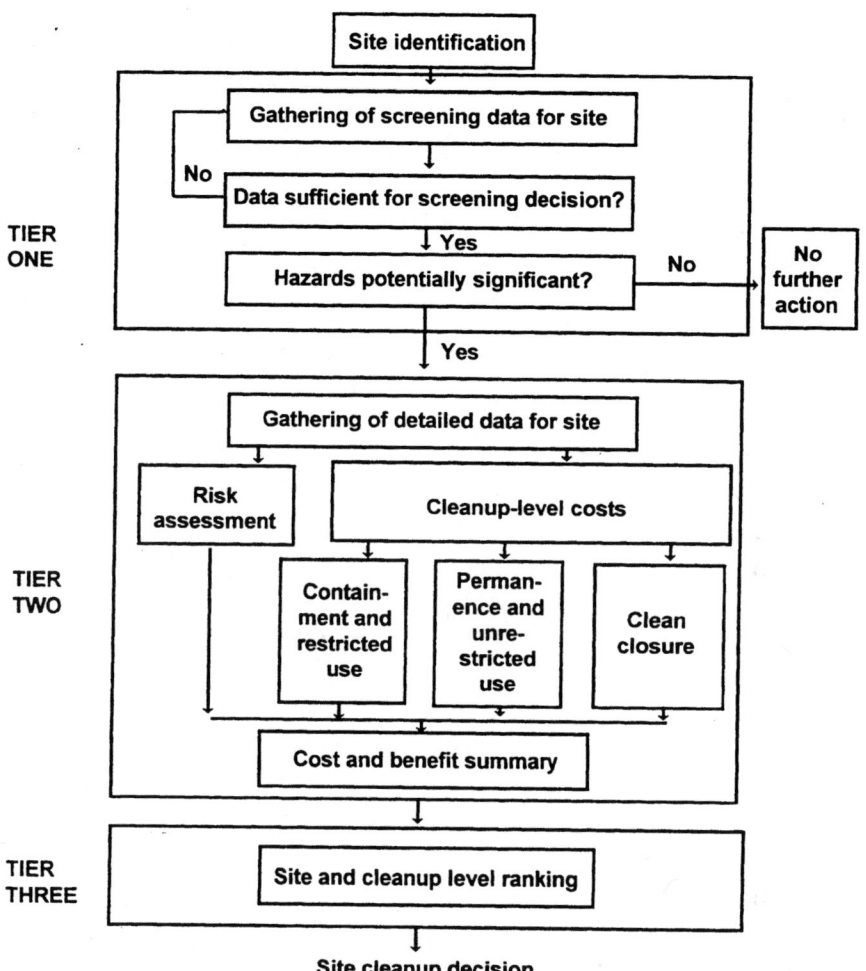

FIGURE 9-1. General outline of three-tiered approach to a unified process for setting priorities. A continuing site-monitoring program would be needed for discovering new sites and evaluating changes over time in the potential hazards of discovered sites not undergoing remediation.

is to obtain, through consensus, a single well-documented and validated model that not only provides a relative ranking of sites based upon risk to human health and the environment, but also denotes the risk reduction achieved by alternative levels of remediation. The model to be used for this purpose should be tested with validation studies at actual sites, and the uncertainty in the model output should be characterized so that the degree of confidence in the site rankings and the rankings of effectiveness for alternative levels of remediation can be estimated and considered. As an example, see the discussion of such analysis in Chapter 5. Such a model could also help to track progress during remediation.

The site data obtained under Tier Two should be adequate to perform a cost evaluation for site remediation. Here, present value costs would be estimated for each of three levels of remediation. The first level would involve sufficient remediation to contain the hazardous contaminants so that they would not present significant risk to human health and the environment. A no-action alternative might be equivalent to this level at some sites; at other sites, all contaminants would not necessarily be removed, and land use controls and restricted access might be required. The second level of remediation considered would restore the site to the point where no land use restrictions would be necessary. This level of cleanup would be equivalent to "permanence," in that no continuing costs for contaminant containment would be required. The third level of cleanup would be more extensive, comparable to returning the site to precontamination quality. Present value costs to achieve each of these three levels of control would be estimated. The three levels of control suggested here are similar to the three levels considered by Russell et al. (1991) in their evaluation of costs to the nation of hazardous waste remediation.

In Tier Three, a ranking resulting from the Tier Two assessment of risk to human health and the environment, together with the cost estimated for each of the three levels of control, would be determined by an independent process. By giving a ranking to the site, a determination could be made regarding what sites to address first and what levels of control to instigate. The committee does not recommend a particular framework for doing this, but clearly one is needed. This process might involve some mathematical formalization, but that should be coupled with broader political, social, and economic considerations. Decisions at this level should be made in a well defined process. The ranking process could be centralized or decentralized, and individuals or groups who have an explicit objective in mind might contribute to the process. Today, such decisions on ranking and site cleanup level are generally made by an informal and ill-defined process that is peculiar to the agency involved. The process used needs to be more explicit than is the current practice, so that national resources can be expended in a more open and cost-effective manner.

There are several advantages to such a three-tiered unified national scientific approach. First, this process for priority setting proposed is similar overall to approaches currently being used among federal and state agencies. Thus, no radical change in thinking is required. It does not involve any greater degree of complexity than currently used procedures, and indeed would take advantage of knowledge gained from application of current processes. The main advantage is that the same procedure would be used in Tier One and Tier Two by all agencies. This would allow greater effort to be placed in evaluating and improving the scientific basis for the mathematical procedures being used and would be more cost efficient for determining how the models perform with respect to their intended purposes, for evaluating the validity of the approach used, and for determining the sensitivity of the model to data inputs. The overall cost for developing good model documentation and acquiring appropriate input coefficients would be reduced. The steps required for model development and validation can therefore be more readily implemented in a unified approach, so that the result can attain the credibility necessary for broad confidence in its use.

The three levels of cost for site remediation distinguished under Tier Two would allow better judgments to be made concerning the degree of cleanup that should be pursued at a given site. For example, suppose for a given site (Site One), the estimated cost for the first level of cleanup is $2 million and for the second level it is $3 million. For a second site (Site Two), the estimated cost for the first level of remediation is $2 million and for the second level it is $100 million. If the risks posed by the two sites are similar and funds available for remediation are limited, it would be understandable and reasonable to suggest that Site One be cleaned at least to the second level, while the first level of cleanup may be the best alternative for Site Two.

The benefits of cleanup would also be calculated. The decision for the two sites in the example above might change if the benefits of the second level remediation of Site Two were $400 million compared to $100 million in costs.

Assuming that all three levels of cleanup would be sufficiently protective of human health and the environment, the degree of permanence and the extent of land use restrictions might be quite different. As with any priority-setting approach, a continuing program would be needed for discovering new sites and for evaluating changes over time in the potential indicated hazards from discovered sites not undergoing remediation. Such a monitoring program would require efforts commensurate with the potential hazard of specific waste sites. For example, closer monitoring of "no-action" sites and "containment-only" sites would be needed compared with monitoring unlisted sites. Such a program would be greatly facilitated by a uniformly-applied and validated model.

Because the approach proposed by the committee is not dependent on a certain period, it could be used to incorporate considerations of long-term risk. For example, longer periods could be used in the analysis performed under Tier Two of the recommended approach where risk is assessed for each site. Also, sites rejected from further consideration in Tier One could be fed back into the process periodically. However, how long a site should be monitored and assessed for risk and the timing of remedial actions are issues that are beyond the charge of this committee, but should be the subject of a separate study. OTA (1989) discussed some of the benefits and concerns regarding a priority-setting process that explicitly addresses future risk.

Under the proposed three-tiered system, hazards posed by various sites and the relative costs for different degrees of cleanup would be provided explicitly, and in a manner that is understandable by the public and decision-makers. This would make it easier for decision-makers and affected parties, both public and private, to arrive at rational decisions for setting priorities and levels for cleanup of contaminated sites. If used as a rational process for isolating and elucidating data limitations and scientific uncertainties, and for articulating and implementing societal values, it will enhance our ability to make prudent decisions.

10

CONCLUSIONS AND RECOMMENDATIONS

NEED FOR PRIORITY-SETTING PROCESS

The enormous costs and technical limits to cleaning up hazardous-waste sites highlight the need for a more comprehensive, explicit, and systematic approach to setting cleanup priorities.

From a review of the hazardous-waste cleanup problem in the United States, and the technological limits and enormous costs to reaching the original goals for cleanup stipulated by Congress, it is apparent that the original assumption of a few sites needing remediation was incorrect. It is also apparent that there are tens of thousands of sites potentially costing hundreds of billions of dollars to cleanup. Thus, it no longer suffices to have ranking models that only attempt to identify the bad sites. There are too many of them. Faced with this reality, society needs to take the next step and develop an overall priority-setting system that helps define a more comprehensive, explicit, and systematic cleanup strategy, addressing such questions as where the available funds should be spent, and how they should be spent. Toward this end, the committee reviewed and compared the current models for ranking hazardous-waste sites and the overall priority-setting systems being used or proposed by various federal agencies and the states. An approach toward a unified national priority setting system was developed. Conclusions and recommendations reached from the committee's efforts are contained in the following.

The committee's general conclusions and recommendations are presented below: first, for the overall priority-setting process for remediation of contaminated sites and, second, for the mathematical models that are used as part of the overall process for ranking sites. Conclusions and recommendations for the priority-setting processes and ranking models of specific agencies are presented in earlier chapters.

CURRENT PRIORITY-SETTING

The current priority-setting processes for hazardous-waste site cleanup are not well defined and appear to lack adequate evaluation, sufficient consistency, and effective oversight.

Confronting the great number of waste sites and potential hazards that have been identified, EPA, DOE, and DOD had to develop ranking and priority-setting systems for remediation. The missions or mandates for these systems are diverse and complex. The scope and scale of the national effort to remediate waste sites

CONCLUSIONS AND RECOMMENDATIONS

in the private and public sectors is much larger than was conceived by any of the agencies. Resource requirements parallel those of other major societal activities and challenges such as national debt reduction, the savings and loan crisis, and infrastructure renewal. Setting priorities for remediating the sites requires a well-organized and well-defined national approach and commitment. This has not yet been achieved.

None of the agencies have developed its overall priority-setting process in a manner that is explicit, adequately documented, and sufficiently open to scientific and public scrutiny.

The overall process for setting priorities for remediation of hazhazardous-ardous waste sites was found generally to involve three major tiers of activity: (1) site screening to determine which sites will receive a detailed evaluation for further decisions; (2) preparing a detailed technical evaluation of the situation at each site chosen in Step 1; and (3) setting of priorities and procedures for remediation. Most federal and state agencies follow an overall process of setting priorities for site remediation that includes the three phases in some manner. However, much of that process remains opaque and thus potentially lacks credibility.

In considering the procedure used by EPA, DOD and DOE, the committee has determined that many of the steps in determining priorities are external to the ranking models that serve as aids in the process, have not been explicitly articulated, and therefore remain obscured from public scrutiny. The DOD and DOE priority-setting processes for their own sites have inadequate independent oversight. Such a process leaves the polluter itself (DOD or DOE) responsible for discovering the pollution, investigating and characterizing the extent of pollution, selecting remedial approaches, setting priorities for remediation of sites, and executing and monitoring the remediation, a situation which cannot help but undermine the credibility of the process.

Ranking Hazardous Waste Sites

There is no consolidated ranking of sites at the national level.

The federal agencies with hazardous-waste sites needing remediation are approaching ranking and setting priorities of sites differently. The ranking and priority-setting processes used by DOD, DOE, and EPA have been considered in detail in this report, but other agencies have different approaches as well. The other agencies include the Department of the Interior, the Department of Transportation, the Small Business Administration, and the National Aeronautics and Space Administration, all of which have hazardous-waste sites on the NPL, and other agencies that have sites on EPA's Federal Facility Docket. Many contaminated sites are not included in the present ranking processes because they are under different programs. An example of this is the exclusion of sites and facilities covered by the Resource Conservation and Recovery Act. Other examples include DOD sites for storing and destruction of chemical and biological weapons (demilitarization program); sites covered by DOD's Defense Environmental Restoration Program; DOD's Base Realignment and Closure sites; DOE's Environmental Restoration Program sites and Waste Management sites, as well as other DOE hazardous-waste sites not covered by either of these programs, e.g., sites under FUSRAP (Formerly Utilized Sites Remedial Action Program).

IMPROVING THE PRIORITY-SETTING PROCESS

To the maximum extent possible, the overall priority-setting processes, including the mathematical models used, should be similar across the various federal agencies.

Escalated remediation costs (actual and projected) and insufficient public accountability of desperate efforts suggest that the existing assortment of processes for priority setting should be

CONCLUSIONS AND RECOMMENDATIONS

drastically changed. A unified national process—based on the use of similar information and models at every site and is consistent for all states and areas within states—is recommended.

In view of the enormous direct public investment required for remediation of contaminated federal facilities and indirect payment for remediation at private sites, a single well-developed and documented process is needed that not only ensures use of funds on a consistent basis and a proper return to the nation from this investment but also engenders confidence within the scientific community and the public.

This uniform national priority-setting process should be more scientifically based, explicit, and open and accessible to the public than has been the case for the three major tiers of the overall priority-setting process discussed in Chapter 9. Openness is at least as important as scientific validity. The complete priority-setting process should be well documented, it should be subject to review by the scientific community and the affected public, and it should explicitly address not only risks to human health and the environment, but also social and economic issues.

This consistency and openness should apply to everything from data requirements to requirements for addressing social, economic, and cultural factors. A specific practical implication is that each tier of activity should be similar in scope and content across the various agencies.

The priority-setting process should have a common mechanism for identifying serious immediate hazards or emergency conditions and pulling them out of the longer-term priority-setting process; all of the systems the committee studied have some such feature. A unified approach should also include a formal site-discovery program, which is currently lacking. It should also include a process for tracking site remediation progress and monitoring sites that may pose dangers far into the future.

Such a national plan or protocol would greatly benefit by a support program, including

- Expanded research and development in the advancement of remediation technologies, analytical methods, knowledge of contaminant movement and fate, methods to measure and estimate health, socioeconomic and environmental impacts, and analyses of institutional barriers retarding the remediation process.
- Technical and scientific education for federal, state, local and private operatives, as well as foreign partners.
- Outreach extension, and technical transfer for the federal, state, private, and public sectors.

CURRENT RANKING MODELS USED IN PRIORITY-SETTING

The formal mathematical models developed to aid in the priority-setting process play little role in determining which sites are ultimately remediated.

Much attention has been given to scores from mathematical models developed for site-ranking, which are used as aids in the priority-setting process. However, they are only one of the factors that are ultimately used to determine whether or not a site will be remediated, the degree to which it will be remediated, and when. Key decisions are made external to the models through negotiations and the political process.

Site-ranking models would play a greater role in the priority-setting process if they incorporated to a greater extent social and economic values, and if users and the public were more confident in model outcomes.

A strong scientific base now exists upon which to build a sound ranking model that could play a larger role in the overall priority-

CONCLUSIONS AND RECOMMENDATIONS

setting process. No large apparent gap exists in this base that would require a major research effort for improvement. This scientific base has already been used to a degree in the development of at least parts of the different ranking models. However, all ranking models were found to fall short on several important attributes of model development, including adequate documentation, proper validation, completeness, transparency, and adequate inclusion of social and economic factors. Sufficient attention to these attributes is necessary for user confidence in model outputs.

The mathematical models used by EPA, DOE, and DOD as aids to setting priorities differ widely.

The three agencies' mathematical models examined by the committee have forms traceable to these agencies individual mandates and to the complexity and number of the sites they manage. EPA needed an early screening model to eliminate the vast majority of nominated sites from further consideration. The major question was what sites should be selected for detailed investigation. DOD needed a screening model for use at a later decision stage when funds for remediation would become inadequate for the need. The question here was when should a given site be remediated. DOE needed a more comprehensive model that could address its fewer but larger facilities, each with many complex sites. The major question was how funds available for remediation of a given site should be distributed among its many contaminated sites in order to optimize reduction of risk to humans and the environment.

The three models have similarities in the environmental pathways they consider, but they differ in the stages within the overall priority-setting process at which they are applied. They differ in the types of input information that they use, in the environmental media they consider for transport of contaminants and exposures, in the relative importance they give to human health and the envi-

273

ronment, and in how they handle social and economic aspects. They also differ greatly in the weighting given to the different exposure pathways, as well as the selection of the factors used to weight the influence of the environmental and toxicological data.

Further, the definition of *site* is not consistently used within any agency, or between different agencies. For example, EPA's HRS has been used to list the Rocky Mountain Arsenal—an entire DOD installation—on the NPL, and also to list Basin F—one small part of the Rocky Mountain Arsenal—on the NPL. Obviously, the risk from many sites represented by the whole Arsenal poses a greater risk than any single site by itself. Also obviously, the cost for cleanup of the entire Arsenal is greater than the cost for any one of its sites. This example suggests why ranking of sites by the risk posed, when remediation costs are not considered, can be quite misleading.

The different models are applied at different steps in the priority-setting process and, as such, have different data and resource needs and provide estimates of site hazard with different levels of accuracy and precision. This might lead to inconsistent rankings between the models, a problem that should be expected and cannot in itself be avoided. However, the relationship between pathway subscores in the three models often differ substantially, as do the weightings given to the different exposure pathways and the selection of factors used to weight the influence of the environmental and toxicological data and the value of natural resources. The different weightings provided often reflect differences in value judgments, but the process by which these value judgments were obtained is not often clear. Because the model results have an effect on the expenditure of vast amounts of public funds, a more consistent and better documented approach to obtaining and using value-laden weighting factors would seem appropriate.

State ranking systems for waste sites follow one of several ap-

CONCLUSIONS AND RECOMMENDATIONS

proaches: a quantitative approach similar to the EPA HRS, other explicit numeric systems leading to a site-specific score, or tend to differentiate all sites into a small number of categories of priority based mainly upon narrative descriptions of site characteristics.

For many of the states considered, there is evidence of very thoughtful development of site ranking models. However, how the relationships between the model parameters were developed and what strategies were used for combining parameters is not always clear and often not documented, thus the models tend to lack credibility. For this reason, similar questions of appropriateness of the logic for combining various scores within the ranking methods applies to states' approaches as well as to approaches of federal agencies.

States not using formal ranking models often tend to develop less data-intensive methods that rely on the judgment of professionals in the state agencies to integrate information into site rankings. A more detailed evaluation, beyond the scope of this study, would be needed to evaluate the utility of such approaches relative to the mathematical modeling methods used by the federal government and other states.

IMPROVING THE MODELS

Ranking models can and should play a greater role in the priority-setting process than is currently the case.

Models can be important tools in a priority-setting process because they can integrate a wide variety of important technical, social, and economic factors. The committee believes that with achievable upgrades in certain aspects, these models could play a more important role in the overall priority-setting process, pointing toward a more equitable distribution of funds and a wiser and more financially sound national effort towards site remediation. Some of these aspects are summarized next.

Documentation and Clarity

An important part of any model development is documentation that permits reviewers to understand why the models are structured the way they are and the process by which coefficients that reflect value judgments have been derived. In addition, although models may be technically sophisticated, their core elements should, to the maximum extent feasible, be intuitively as clear as possible to technical and nontechnical audiences. These aspects are important to increase the understanding of the model or process and the acceptance of the results produced.

The output from a ranking model should provide information in addition to the overall score itself so that one can understand why a high or low score was obtained. The additional information would include individual environmental pathway scores, whether site contaminants pose acute or chronic risks, and how the model's value-weights affect the overall score.

Public Involvement

The process of developing a model (or any major component of the model) should be as open as possible, involving both stakeholders and the technical community. Value preferences should be explicit in the models, and coefficients reflecting these preferences should be developed with the affected parties in an open and well-defined process. The process of applying the model to a given

Conclusions and Recommendations

site (or to a large installation such as a military base or a DOE facility) should be similarly open, so that there is the greatest understanding of the results of the model.

Validation

The development and introduction of any important decision-aiding model, such as those under discussion here, should include an explicit process for validating the components of the model and the overall model itself. Flexibility should be provided for revising the components of the model to reflect new knowledge. Adequate validation to objective criteria might require the development of a collection of test sites that have agreed-upon priority rankings, resulting from a comprehensive evaluation, against which to compare the results of model output. The purpose is to test model results and to build user confidence in model outcomes.

Although it is not validation in the strict sense, comparing the performance of one ranking model with performances of other ranking models is a useful exercise (see Chapter 8). The approach should be used on a regular basis in the future to compare the performance of newly developed or modified models against the output of established ones. A set of reference sites should be established for use in developing input data for a wide range of hazardous potential (high to low). Using a set of about 12 dissimilar sites would help ensure that the models are compared on the basis of a broad range of site characteristics. The approach would be useful for checking whether revised algorithms are performing as expected or whether the models can discriminate, in a numerical sense, among various degrees of potential site hazards in the range of high to low.

Explicit Consideration of Socioeconomic Effects

Consideration of danger to public welfare by EPA is required under CERCLA, and part of the Agency for Toxic Substances and Disease Registry's mandate in SARA is to consider negative effects (of waste-site contaminants) on quality of life. A comprehensive site evaluation model should include explicit considerations not only of human health and the environment, but also of socioeconomic impacts on the surrounding community. Such considerations are probably always part of the priority-setting process, but they generally are not made explicitly, and so are not open to public scrutiny and evaluation. If this important element in setting priorities can be given a common explicit basis, then greater confidence in the overall process will be achieved. Methodologies that allow the incorporation of rigorous socioeconomic impact assessments directly into models for ranking hazardous sites are currently available. Omission of an explicit treatment of these socioeconomic components in hazard ranking models can lead to a biased priority-setting agenda.

REFERENCES

Advisory Committee on Nuclear Facility Safety to the U.S. Department of Energy. 1991. Final Report on DOE Nuclear Facilities. U.S. Department of Energy, Washington, D.C.

Anderson, A., and P. Couture. 1984. The U.S. Army Installation Restoration Program. Pp. 511-516 in the 5th National Conference on Management of Uncontrolled Hazardous Waste Sites, Nov. 7-9.

Applied Decision Analysis, Inc. 1987. A Site-Ranking Panel Evaluation of the Relative Risk Posed by Twenty Superfund Sites. Draft report to the Office of Policy and Planning Evaluation, U.S. Environmental Protection Agency, Washington, D.C.

Arcos, J.C., Y-t. Woo, and G. Polansky. 1989. Ranking of complex mixtures for potential cancer hazard: Structure of a computerized system—An outline. J. Environ. Sci. Health C7:129-144.

Arula, X.J.R. 1987. Mathematical modeling. Pp. 719–728 in Encyclopedia of Physical Science and Technology, Vol. 7. New York: Academic Press.

ASTM (American Society for Testing and Materials). 1984. Practice for Evaluating Environmental Fate Models of Chemicals. Designation E978-84, Committee on Biological Effects and Environmental Fate. American Society for Testing and Materials, Philadelphia, Pa.

Barnthouse, L.W., J.E. Breck, T.D. Jones, S.R. Kraemer, E.D. Smith, and G.W. Suter II. 1986. Development and Demonstration of a Hazard Assessment Rating Methodology for Phase II of the Installation Restoration Program. ORNL/TM-9857. Oak Ridge National Laboratory, Oak Ridge, Tenn.

Brown, K.W., J.F. Slowey, and H.W. Wolf. 1977. Accumulation and Passage of Pollutants in Domestic Septic Tank Disposal Fields. R801955-01-2. Final Report to U.S. Environmental Protection Agency, Office of Research and Monitoring, Robert S. Kerr Water Research Center, Ada, Okla.

Butler, J., and W. Jones. 1992. Priorities of Actions on Superfund Sites. Paper presented to the Coalition on Superfund. Putnam, Hayes and Bartlett, Inc., Washington, D.C.

Caldwell, S., and A. Ortiz. 1989. Overview of Proposed Revisions to the Superfund Hazard Ranking System. J. Air Waste Mgmt. Assoc. 39:801-807.

California Department of Toxic Substances Control. 1991. Integrated Site Mitigation Process, Final Draft. California Department of Toxic Substances Control, Sacramento, Calif.

Campbell, C. 1991. Toxins taint bases future. Home News (New Brunswick, N.J.), May 26, p. 1.

Canter, L.W., and L.G. Hill. 1979. Handbook of Variables for Environmental Impact Assessment. Ann Arbor, Mich.: Ann Arbor Science Publishers.

Carpenter, G.F. 1990. Environmental Contamination Site Risk Assessment and Ranking in Michigan: Revision of the Site Assessment System Model. Ph.D. Dissertation. Department of

REFERENCES

Fisheries and Wildlife, Michigan State University, East Lansing, Mich.

CBO (Congressional Budget Office, U.S. Congress). 1990. Federal Liabilities Under Hazardous Waste Laws. U.S. Congress, Congressional Budget Office, Washington, D.C.

Chang, S., K. Barrett, S. Haus, and A. Platt. 1981. Site Ranking Model for Determining Remedial Action Priorities Among Uncontrolled Hazardous Substances Facilities. Mitre Corporation, McLean, Va.

Chapra, S.C., and K.H. Reckhow. 1983. Engineering Approaches for Lake Management, Vol. 2. Mechanistic Modeling. Boston: Butterworth Publishers.

Conservation Foundation. 1987. State of the Environment: A View Toward the Nineties. Washington, D.C.: Conservation Foundation.

Cohrssen, J.J., and V.T. Covello. 1989. Risk Analysis: A Guide to Principles and Methods for Analyzing Health and Environmental Risks. U.S. Council on Environmental Quality, Executive Office of the President, Washington, D.C.

Crouch, E.A.C. and R. Wilson. 1981. The regulation of carcinogens. Risk Anal. 1:47-57.

Crouch, E.A.C., R. Wilson, and L. Zeise. 1983. The risks of drinking water. Water Resources Res. 19:1359-1375.

Davos, C.A. 1977. Priority-tradeoff-scanning approach to evaluation in environmental management. J. Environ. Mgmt. 5:259-273.

DOD (U.S. Department of Defense). 1987. Directive, Real Property Acquisition, Management, and Disposal. U.S. Department of Defense, Washington, D.C.

DOD (U.S. Department of Defense). 1990a. Defense Priority Model, User's Manual, Version 3.0. Office of the Deputy Assistant Secretary of Defense, U.S. Department of Defense, Washington, D.C.

DOD (U.S. Department of Defense). 1990b. Defense Priority

Model FY 90 Scoring Quality Assurance Program Report. Report submitted to the Defense Environmental Support Office, Office of the Deputy Assistant Secretary of Defense (Environment) by Peer Consultants, Rockville, Md.

DOD (Department of Defense). 1991a. Installation Restoration Program Cost Estimate. Office of the Deputy Assistant Secretary of Defense (Environment), U.S. Department of Defense, Washington, D.C.

DOD (U.S. Department of Defense). 1991b. Defense Priority Model, User's Manual, FY 92 Version. Office of the Deputy Assistant Secretary of Defense (Environment), U.S. Department of Defense, Washington, D.C.

DOE (U.S. Department of Energy). 1990. A Premliminary Conceptual Design of a Formal Priority System for Environmental Restoration. Office of Environmental Restoration and Waste Management, U.S. Department of Energy, Washington, D.C.

DOE (U.S. Department of Energy). 1991a. Preliminary Design Report. DOE Environmental Restoration Priority System. Office of Environmental Restoration and Waste Management, U.S. Department of Energy, Washington, D.C.

DOE (U.S. Department of Energy). 1991b. Instructions for Generating Inputs for the FY 1993 Application of the DOE Environmental Restoration Priority System. Office of Environmental Restoration, U.S. Department of Energy, Washington, D.C.

Donahue, R.L., R.W. Miller, and J.C. Shickluna. 1977. Soils: An Introduction to Soils and Plant Growth, 4th ed. Englewood Cliffs, N.J.: Prentice-Hall.

Doty, C.B., and C.C. Travis. 1990. Is EPA's national priorities list correct? Environ. Sci. Technol. 24:1778-1780.

Dyer, J.S. 1990. Remarks on the analytic hierarchy process. Mgmt. Sci. 36:249-258.

Edelstein, M. 1988. Contaminated Communities: The Social and Psychological Impacts of Residential Toxic Exposure. Boulder, Colo: Westview Press.

REFERENCES

Edwards, W., and J.R. Newman. 1982. Multi-attribute Evaluation. Quantitative Applications in the Social Sciences, Vol. 26. Troy, N.Y.: Sage Publications.

ELI (Environmental Law Institute). 1989. An Analysis of State Superfund Programs: A 50-State Study. EPA/540/8-89/011. Prepared for the U.S. Environmental Protection Agency, Contract 68-W8-0098, by the Environmental Law Institute, Washington, D.C.

ELI (Environmental Law Institute). 1991. An Analysis of State Superfund Programs: 50-State Study, 1991 Update. Prepared for the U.S. Environmental Protection Agency, Contract CR-817553-01, by the Environmental Law Institute, Washington, D.C.

Ellenhorn, M.J., and D.G. Barceloux. 1988. Medical Toxicology: Diagnosis and Treatment of Human Poisoning. New York: Elsevier.

Environmental Response Division. 1990. User's Manual Site Assessment Model. Environmental Response Division, Michigan Department of Natural Resources, Lansing, Mich.

EPA (U.S. Environmental Protection Agency). 1977. Report to Congress: Waste Disposal Practices and Their Effects on Ground Water. U.S. Environmental Protection Agency, Washington, D.C.

EPA (U.S. Environmental Protection Agency). 1988. Review of the Superfund Hazard Ranking System. SAB-EC-88-008. Review by the Hazard Ranking System Review Subcommittee of the Science Advisory Board. Office of the Administrator, Science Advisory Board, U.S. Environmental Protection Agency, Washington, D.C.

EPA (U.S. Environmental Protection Agency). 1989a. A Management Review of the Superfund Program. U.S. Environmental Protection Agency, Washington, D.C.

EPA (U.S. Environmental Protection Agency). 1989b. Methods for Evaluating the Attainment of Cleanup Standards, Vol. 1:

Soils and Solid Media. EPA/230/2-89-042. Office of Policy, Planning, and Evaluation, U.S. Environmental Protection Agency, Washington, D.C.

EPA (U.S. Environmental Protection Agency). 1989a. Risk Assessment Guidance for Superfund, Vol. 1: Human Health Evaluation Manual, Part A, Baseline Risk Assessment. Interim Final. EPA/540/1-89-002. Office of Emergency and Remedial Response, U.S. Environmental Protection Agency, Washington, D.C.

EPA (U.S. Environmental Protection Agency). 1989b. Soil Sampling Quality Assurance User's Guide, 2nd ed. EPA/600/8-89-046. Environmental Monitoring Systems Laboratory, U.S. Environmental Protection Agency, Las Vegas, Nev.

EPA (U.S. Environmental Protection Agency). 1990a. Environmental Investments: The Costs of a Clean Environment. U.S. Environmental Protection Agency, Washington, D.C.

EPA (U.S. Environmental Protection Agency). 1990b. Superfund: Focusing on the Nation at Large. U.S. Environmental Protection Agency, Washington, D.C.

EPA (U.S. Environmental Protection Agency). 1990c. Final Policy on Setting RI/FS Priorities. OSWER Directive No. 9200.3-11. Office of Solid Waste and Emergency Response, U.S. Environmental Protection Agency, Washington, D.C.

EPA (U.S. Environmental Protection Agency). 1991a. National Priorities List, Supplementary Lists and Supporting Materials. Office of Emergency and Remedial Response, U.S. Environmental Protection Agency, Washington, D.C.

EPA (U.S. Environmental Protection Agency). 1991b. Office of Emergency and Remedial Response Publication 9345.0-06. U.S. Environmental Protection Agency, Washington, D.C.

EPA (U.S. Environmental Protection Agency). 1991c. PA Method: Development and Test Results. U.S. Environmental Protection Agency, Washington, D.C.

REFERENCES

Federal Facility Agreement. In the Matter of the U.S. Department of the Air Force, Air Force Bases.

Federal Facilities Environmental Restoration Dialogue Committee. 1993. Recommendations for Improving the Federal Facility Environmental Restoration Decision-Making Process and Setting Priorities in the Event of Funding Shortfalls. U.S. Environmental Protection Agency, Washington, D.C.

Federal Register. 1981. Presidential Documents, Executive Order 12316. Responses to Environmental Damage. Fed. Regist. 46(161):42237-42240.

Federal Register. 1989. Defense Priority Model: Defense Environmental Restoration Program. Action: Notice of Plans to Implement. Fed. Regist. 54(202):43104-43106.

Federal Register. 1990. Hazard Ranking System: Final Rule. Fed. Regist. 55(241):51532-51667.

Federal Register. 1992. Guidelines for Exposure Assessment. Fed. Regist. 57(104):22888-22938.

Foley, R.E., S.J. Jackling, R.J. Sloan, and M.K. Brown. 1988. Organochlorine and mercury residues in wild mink and otter: Comparison with fish. Environ. Toxicol. Chem. 7:363-374.

Freeze, R.A., and J.A. Cherry. 1979. Groundwater. Englewood Cliffs, N.J.: Prentice-Hall.

GAO (General Accounting Office). 1987. Simulations: Improved Assessment Procedures Would Increase the Credibility of Results. GAO/PEMD-88-3. U.S. General Accounting Office, Washington, D.C.

GAO (U.S. General Accounting Office). 1991. Hazardous Waste: DOD Estimates for Cleaning Up Contaminated Sites Improved but Still Constrained. GAO NSIAD 92-37. U.S. General Accounting Office, Washington, D.C.

Gass, S.I. 1983. Decision-aiding models: Validation, assessment, and related issues for policy analysis. Comput. Ops. Res. 31:603-631.

Gass, S.I., and B.W. Thompson. 1980. Guidelines for model evaluation: An abridged version of the U.S. General Accounting Office Exposure Draft. Comput. Ops. Res. 28:431-439.

Gass, S.I., and L.S. Joel. 1981. Concepts of model confidence. Comput. Ops. Res. 8:341–346.

Golden, B.L., E.A. Wasil, P.T. Harker, and J.M. Alexander, eds. 1989. The Analytic Hierarchy Process: Applications and Studies. New York: Springer-Verlag.

Greenberg, M., and R. Anderson. 1984. Hazardous Waste Sites: The Credibility Gap. New Brunswick, N.J.: Center for Urban Policy Research.

Griffin, R.A., B.L. Herzog, T.M. Johnson, W.J. Morse, R.E. Hughes, S.F.J. Chou, and L.R. Follmer. 1985. Mechanisms of contaminant migration through a clay barrier—Case study, Wilsonville, Illinois. Pp. 27-38 in Land Disposal of Hazardous Waste. Proceedings of the Eleventh Annual Research Symposium. EPA 600/9-85/013. Office of Research and Development, U.S. Environmental Protection Agency, Cincinnati, Ohio.

Halfon, E. 1989. Comparison of an index function and a vectorial approach method for ranking waste disposal sites. Environ. Sci. Technol. 23:600-609.

Halfon, E., and M.G. Reggiani. 1986. On ranking chemicals for environmental hazard. Environ. Sci. Technol. 20:1173-1179.

Hall, C.W. 1988. Practical limits to pump and treat technology for aquifer remediation. Pp. 7-12 in Proceedings: Groundwater Quality Protection Pre-Conference Workshop. Water Pollution Control Federation, Washington, D.C.

Hammer, P.L., Keeney, R.L., R.H. Möhring, H. Otway, F.J. Radermacher, and M.M. Richter, eds. 1988. Multi-attribute Decision Making via O.R.-based Expert Systems. Proceedings of the International Conference on Multi-Attribute Decision Making via O.R.-based Expert Systems, April 20-27. Ann. Operat. Res. 16.

REFERENCES

Haness, S.J., and J.J. Warwick. 1991. Evaluating the hazard ranking system. J. Environ. Mgmt. 32:165-176.

Harker, P.T., and L.G. Vargas. 1990. Reply to Remarks on the Analytic Hierarchy Process by J.S. Dyer. Mgmt. Sci., pp. 269-273.

Harris, R.H., and G.C. Wrenn. 1988. Making Superfund work. Issues Sci. Technol. 4(3):54-58.

Hill, M. 1968. A goals-achievement matrix for evaluating alternative plans. J. Am. Inst. Planners 34:19-29.

Hird, J.A. 1990. Superfund expenditures and cleanup priorities: Distributive politics and the public interest? J. Policy Anal. Mgmt. 9:455-483.

Home News. 1991. Bill would stop Superfund polluters from passing buck. Home News (New Brunswick, N.J.), July 19, p. A4.

Hwang, C.L., and K. Yoon, K. 1981. Multiple attribute decision making: Methods and applications: A state-of-the-art survey. Lecture Notes in Economics and Mathematical Systems, Vol. 186. New York: Springer-Verlag.

Hyman, E.L., and B. Stiftel. 1988. Combining Facts and Values in Environmental Impact Assessment: Theories and Techniques. Boulder, Colo.: Westview Press.

Inhaber, H. 1976. Environmental Indices. New York: John Wiley-Interscience.

Inside E.P.A. 1991. EPA officials call for major overhaul to standardize Superfund cleanups. Inside E.P.A. Weekly Rep. 12(39):Sept. 27.

Inside E.P.A. 1992. Reilly approves plan to speed superfund waste removal, reduce study time. Inside E.P.A. Weekly Rep. 13(10):March 6.

Johnson, R.C., and L.J. Zaragoza. 1991. The Hazard Ranking System and Relative Risk. Paper presented at a meeting of the Division of Environmental Chemistry, American Chemical Society, Aug. 25-30, New York.

Julien, B., S.J. Fenves, and M.J. Small. 1992. Knowledge acquisition methods for environmental evaluation. AI Appl. 6:1-20.

Keeney, R.L. 1977. The art of assessing multi-attribute utility functions. Organiz. Behav. Hum. Perf. 19:267-310.

Keeney, R.L., and H.R. Raiffa. 1976. Decisions with Multiple Objectives: Preferences and Value Tradeoffs. New York: John Wiley & Sons.

Keystone Center. 1991. Keystone National Dialogue on Federal Facility Environmental Management. First Plenary Session Summary, June 12-13, Washington, D.C.

Krupnick, A.J., and P.R. Portney. 1991. Controlling urban air pollution: A benefit-cost assessment. Science 252:522-527.

Landy, M.K., M.J. Roberts, and S.R. Thomas. 1990. The Environmental Protection Agency: Asking the Wrong Questions. New York: Oxford University Press.

Lave, L., and H. Gruenspecht. 1991. Increasing the efficiency and effectivensss of environmental decisions: Benefit-cost analysis and effluent fees. J. Air Waste Manage. Assoc. 41:680-693.

Lichfield, N., P. Kettle, and M. Whitbread. 1975. Evaluation in the Planning Process. Oxford, U.K.: Pergamon Press.

Mackay, D.M., and J.A. Cherry. 1989. Groundwater contamination: Pump-and treat remediation. Environ. Sci. Technol. 23(6):630-636.

Mazmanian, D., and D. Morell. 1992. Beyond Superfailure. Boulder, Colo.: Westview Press.

McAllister, D.M. 1980. Evaluation in Environmental Planning: Assessing Environmental, Social, Economic and Political Trade-Offs. Cambridge, Mass.: MIT Press.

McClelland, G., W. Schulze, and B. Hurd. 1990. The effect of risk beliefs on property values: A case study of a hazardous waste site. Risk Anal. 10:485-497.

Michel, K.L. 1992. An Overview of the Multimedia Environmental Pollutant Assessment System. ES/ER/TM-14. Environmental Restoration Division, Oak Ridge, Tenn.

REFERENCES

Missouri Department of Natural Resources. 1991. Confirmed Abandoned or Uncontrolled Hazardous Waste Disposal Sites in Missouri and Hazardous Waste Remedial Fund Statement of Revenues, Expenditures, and Changes in Fund Balance. Annual Report. Division of Environmental Quality, Missouri Department of Natural Resources, Jefferson City, Mo.

MIT (Massachusetts Institute of Technology). 1992. Breaking the Backlog: Improving Superfund Priority Setting. Paper prepared for the Coalition on Superfund. The Center for Technology, Policy and Industrial Development, Massachusetts Institute of Technology, Cambridge, Mass.

Montana Department of Health and Environmental Sciences. 1991. Solid and Hazardous Waste Bureau Superfund Program CECRA Facility Ranking Process. Solid and Hazardous Waste Bureau, Montana Department of Health and Environmental Sciences, Helena, Mont.

Naylor, T.H., and J.M. Finger. 1967. Verification of computer simulation models. Mgmt. Sci. 14:92-101.

Neuhold, J., and L. Ruggiero. 1976. Ecosystem Processes and Organic Contaminants. NSF/RA 76-0008. National Science Foundation, Washington, D.C.

New York State Department of Environmental Conservation. 1992. Inactive Hazardous Waste Disposal Sites in New York State, Annual Report. Division of Hazardous Waste Remediation, New York State Department of Environmental Conservation, Albany, New York.

NJDEPE (New Jersey Department of Environmental Protection and Energy). 1992. Proposed Rule: Cleanup Standards for Contaminated Sites. New Jersey Register (24 N.J.R. 373):Feb. 3.

NRC (National Research Council). 1983. Risk Assessment in the Federal Government: Managing the Process. Washington, D.C.: National Academy Press.

NRC (National Research Council). 1989. Improving Risk Communication. Washington, D.C.: National Academy Press.

NRC (National Research Council). 1990a. Ground Water and Soil Contamination Remediation: Toward Compatible Science, Policy, and Public Perception. Washington, D.C.: National Academy Press.

NRC (National Research Council). 1990. Ground Water Models: Scientific and Regulatory Applications. Washington, D.C.: National Academy Press.

NRC (National Research Council). 1991. Human Exposure Assessment for Airborne Pollutants: Advances and Opportunities. Washington, D.C.: National Academy Press.

NRC (National Research Council). 1992. The Department of Defense Priority Model for Hazardous Waste Site Restoration: An Independent Assessment of Methods, Assumptions, and Constraints. Interim Report. Washington, D.C.: National Academy Press.

NRC (National Research Council). 1993. Issues in Risk Assessment. Washington, D.C.: National Academy Press.

NRC (National Research Council). 1994a. Science and Judgment in Risk Assessment. Washington, D.C.: National Academy Press.

NRC (National Research Council). 1994b. Building Consensus Through Risk Assessment and Management of the Department of Energy's Environmental Remediation Program. Washington, D.C.: National Academy Press.

Odell, R. 1976. NEPA—It's still a hard act to follow. Environ. Act. 8(July 31):4-8.

Ohio Division of Emergency and Remedial Response. 1992. Ohio Ranking Method Manual. (Under revision). Division of Emergency and Remedial Response, State of Ohio Environmental Protection Agency, Columbus, Ohio.

Oregon Department of Environmental Quality. 1991. Appendix A of Inventory Ranking Rule: Site Scoring Procedures. Envi-

REFERENCES

ronmental Cleanup Division, Oregon Department of Environmental Quality, Portland, Oreg.

OTA (Office of Technology Assessment, U.S. Congress). 1985. Superfund Strategy. OTA-ITE-252. Washington, D.C.: U.S. Government Printing Office.

OTA (Office of Technology Assessment, U.S. Congress). 1989. Coming Clean: Superfund's Problems Can Be Solved. OTA-ITE-433. Washington, D.C.: U.S. Government Printing Office.

OTA (Office of Technology Assessment, U.S. Congress). 1991. Complex Cleanup: The Environmental Legacy of Nuclear Weapons Production. OTA-O-485. Washington, D.C.: U.S. Government Printing Office.

Passell, P. 1991. Experts question staggering costs of toxic cleanups. New York Times, Sept. 1, p. 1.

Plaster, E.J. 1985. P. 393 in Soil Science and Management. Albany, N.Y.: Delmar Publishers.

Prest, A.R., and R. Turvey, eds. 1965. Cost-benefit analysis. Pp. 155-207 in Surveys of Economic Theory: Resource Analysis, American Economic Association and the Royal Economics Society. New York: St. Martin's Press.

Puskin, J.S. 1992. An analysis of the uncertainties in estimates of radon induced lung cancer. Risk Anal. 12:277-285.

Reckhow, K.H., J.T. Clements, and R.C. Dodd. 1990. Statistical evaluation of mechanistic water-quality models. J. Environ. Eng. 116:250-268.

Rezendes, V. 1992. Nuclear Weapons Complex: Improving DOE's Management of the Environmental Cleanup. Statement of V. Rezendes before the Department of Energy Defense Nuclear Facilities Panel Committee on Armed Services, House of Representatives. GAO/T-RCED-92-43. U.S. General Accounting Office, Washington, D.C.

Reid, R.C., J.M. Prausnitz, and T.K. Sherwood. 1977. The Properties of Gases and Liquids, 3rd ed. New York: McGraw-Hill.

Ridley, S. 1987. State of the States. Washington, D.C.: Renew America.

Roberts, L. 1991. Costs of a clean environment. Science 251: 1182.

Roy, W.R., and R.A. Griffin. 1989. In-situ Extraction of Organic Vapor from Unsaturated Porous Media. Open File Report prepared for the Environmental Institute for Waste Management Studies, University of Alabama, Tuscaloosa, Ala.

Russell, M., E.W. Colglazier, and M.R. English. 1991. Hazardous Waste Remediation: The Task Ahead. Waste Management Research and Education Institute, University of Tennessee, Knoxville, Tenn.

Ryan, A.P., and Y. Cohen. 1989. Chemical transport in the top soil zone: The role of moisture and temperature gradients. J. Hazard. Mater. 22:283-304.

Saaty, T.L. 1980. The Analytic Hierarchy Process. New York: McGraw-Hill.

Saaty, T.L. 1990. An exposition of the AHP in reply to the paper "Remarks on the Analytic Hierarchy Process." Mgmt. Sci. 36:259-268.

Shaeffer, D.L. 1980. A model evaluation methodology applicable to environmental assessment models. Ecol. Model. 8:297-311.

Skaburskis, A. 1989. Impact attenuation in conflict situations: The price effects of a nuisance land-use. Environ. Planning A21:375-383.

Schneider, K. 1991. Military has new strategic goal of cleanup of vast toxic waste. New York Times, Aug. 5, p. 1.

Schleifstein, M. and J. O'Byrne. 1991. The wasteland. The Times-Picayune (New Orleans), March 24, p. 1.

SCS (Soil Conservation Service, U.S. Department of Agriculture). 1972. SCS National Engineering Handbook, Section 4, Hydrology, Suppl. A. Soil Conservation Service, U.S. Department of Agriculture, Washington, D.C.

References

Shulman, S. 1990. Toxic travels: Inside the military's environmental nightmare. Nuclear Times 8:20-32.

Science. 1991. 11.5-million housecleaning. Science 252:643.

Skaburskis, A. 1989. Impact attenuation in conflict situations: The price effects of a nuisance land-use. Environ. Planning A21:375-383.

Smith, K., J. Martin, and E. Als. 1989. Field Investigation to Characterize Relationship Between Groundwater and Subsurface Gas Contamination at a Municipal Landfill. Presented at the Hazardous Materials Control Research Institute's 10th National Conference and Exhibition, Nov. 27-29, 1989, Washington, D.C.

Stafford, R. 1981. Why Superfund was needed. EPA J. 7(June):8-10.

State of California, Toxic Substances Control Program. 1990. Interim Guidance for Preparation of a Preliminary Endangerment Assessment Report. State of California Toxic Substances Control Program, Department of Health Sciences, Health and Welfare Agency, Sacramento, Calif.

Superfund Fact Sheet. 1981. EPA J. 7(June):13.

Technical Review Group of the Department of Energy. 1991. Priority System for Environmental Restoration. U.S. Department of Energy, Washington, D.C.

Thibodeaux, L.J. 1979. Chemodynamics: Environmental Movement of Chemicals in Air, Water, and Soil. New York: John Wiley & Sons.

Thompson, M.A. 1990. Determining impact significance in EIA: A review of 24 methodologies. J. Environ. Mgmt. 30:235-250.

U.S. Congress, House Committee Public Works Transport Hearings. 1983. Hazardous Waste Contamination of Water Resources, Aug. 10. U.S. Department of Defense, Washington, D.C.

Walsh, W., I. Susel, and A. Ronayne. 1991. Setting risk-based priorities—A method for ranking sites for response. Pp. 112-123 in Proceedings of National Research and Development Conference on the Control of Hazardous Materials, Feb. 20-22, Anaheim, Calif.

Washington State Department of Ecology. 1992. Washington Ranking Method: Scoring Manual. Toxics Cleanup Program, Washington State Department of Ecology, Olympia, Wash.

Westman, W.E. 1985. Ecology, Impact Assessment, and Environmental Planning. New York: John Wiley & Sons.

Whelan, G., D.L. Strenge, J.G. Droppo, Jr., B.L. Steelman, and J.W. Buck. 1987. The Remedial Action Priority System (RAPS): Mathematical Formulations. DOE/RL/87-09, PNL 6200. Pacific Northwest Laboratory, Richland, Wash.

Wilson, A.R. 1991. Environmental Risk: Identification and Management. Chelsea, Mich.: Lewis Publishers.

Wood, J.A., and M.L. Potter. 1987. Hazardous pollutants in class II landfills. Hazardous Waste Mgmt. 37:609-615.

Wright, J., and L. Cole. 1985. Risk Management and the Hazardous Waste Problem in State Governments. Rep. No. NSF/PRA-85028. National Science Foundation, Washington, D.C. Available from National Technical Information Service, Springfield, Va. PB 86-187515.

Wu, J.S., and H. Hilger. 1984. Evaluation of EPA's Hazard Ranking System. J. Environ. Eng. 110:797-807.

Zaragoza, L.J. 1990. Cutoff Score Analysis for the Revised Hazard Ranking System (HRS) Docket, U.S. Environmental Protection Agency Memo to HRS Docket. 105NCP-HRS-18-18. U.S. Environmental Protection Agency, Washington, D.C.